EVERYTHING YOU NEED TO KNOW ABOUT EVERYTHING YOU NEED TO KNOW ABOUT THE UNIVERSE

EVERYTHING YOU NEED TO KNOW ABOUT EVERYTHING YOU NEED TO KNOW ABOUT THE UNIVERSE

From the Big Bang to the Big Crunch, in a nutshell

Chris Cooper

PORTICO

Published in the United Kingdom in 2012 by

Portico Books
10 Southcombe Street
London W14 0RA

An imprint of Anova Books Company Ltd

ISBN 978-1-907554-41-4

A CIP catalogue record for this book
is available from the British Library.

10 9 8 7 6 5 4 3 2 1

Designed by Jason Lievesley at Printed Word
Art direction by Blok Graphic, London
Printed and bound by Everbest Printing, China

This book can be ordered direct from the publisher at
www.anovabooks.com

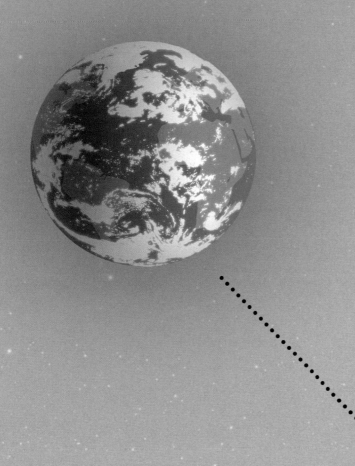

'We are just an advanced breed
of monkeys on a minor planet of
a very average star. But we can
understand the Universe. That makes
us something very special.'

Stephen Hawking

Contents_

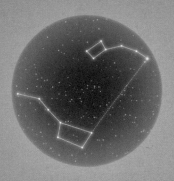

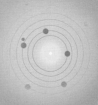

03.0 Probes, Satellites & Spaceships

04.0 The Sun & the Rocky Planets

Contents_continued

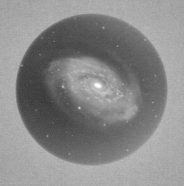

07.0 **Galaxies**

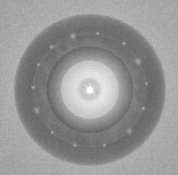

08.0 **The Cosmos**

Introduction_

Space, as Douglas Adams said in *The Hitchhiker's Guide to the Galaxy*, is big – 'really big; vastly, hugely, mind-bogglingly big'. Space is so big that, if you were to spread out all the matter in the Universe evenly, you'd end up with a vacuum that even the most sophisticated laboratory could not match. Nevertheless, that still adds up to an awful lot of stuff to write about. And to make the writer's life even harder, the matter we know about is probably only a fraction of all the matter there is: underneath is a sea of 'dark matter'. We're sure it's there, though we know almost nothing about it – which presumably makes it a known unknown, as Donald Rumsfeld would say.

As well as space, there is also an awful lot of time to contend with. Although the Universe does not have an infinite past (that we know of), it does have 14 billion years' worth of history that we do know about. And unknown billions or trillions of years to come.

And on top of all this there is the possibility that there is an infinite number of universes beyond the observable Universe. So you'll appreciate that quite a lot of compression has been needed to pack the essentials of all this into the small space of this book.

However, the pages that follow will give you some sense of how we and our planet fit into the Universe. You'll learn the essentials of how astronomers – and other scientists – came to their present state of understanding of the Universe: how they extended their senses as their instruments grew more powerful, how they planted their telescopes on mountain tops, in space, and even deep inside mines, and how they learned to 'see' with invisible light. But most of all, you will see that it is the mind's eye rather than the human eye that has enabled astronomers to develop their knowledge, and solve their problems.

One thing is certain: with this bird's-eye view of our knowledge of the cosmos, we – the author, illustrators and publishers – won't have worked ourselves out of a job. As more and more robot spacecraft swarm through the Solar System, as gravity telescopes and neutrino telescopes open new windows on the Universe, and as telescopes are stationed beyond the Moon, this is by no means the end of the story. Although what follows is everything you need to know today, tomorrow there will be more …

Chapter 01.0

The Skies Above Us_

01.1 What's in the Sky?_

The turbulent ocean of air in which we live thins out without any definite boundary, but by one international definition space begins 100km (62 miles) above the Earth's surface. The sky beyond seems to turn around every 24 hours – reflecting the rotation of our planet – and with the naked eye we can see the ever-changing dance of our nearest neighbours, fellow members of the Solar System: the Sun, the Moon, comets and five of Earth's fellow planets. They move against a backdrop of about 6,000 'fixed stars' that maintain the same position in relation to one another. Far beyond these, in the remotest cosmos, we can see just four faint patches that are galaxies beyond our own.

Something to Think About . . .

The stars twinkle simply because of the constantly shifting atmosphere that their light passes through on the way to our eyes. The planets, however, generally do not. Although they look like points of light, the planets actually have discs. Light rays from the different parts of each disc are affected differently by the moving air, and add up to give a steady image. Seen from space or from the surface of the airless Moon, neither planets nor stars twinkle.

* **Aurorae**

  Multi-coloured, shifting, glowing lights in our atmosphere caused by electrically charged particles from the Sun.

* **Meteors**

  Fast-travelling grains of rock that enter the atmosphere from space and burn up from friction. They look like falling stars (hence the name 'shooting star'); some leave a glowing trail that is visible for a second or two.

* **Moon**

  A rocky globe apparently changing shape as we see more or less of its sunlit half.

* **Sun**

  A ball of hot gas that is Earth's nearest star. Very rarely we can see spots on its surface with the naked eye. During an eclipse we can see its extended glowing atmosphere.

* **Planets**

  Five objects that look like stars but move slowly in relation to the background of stars and are much closer to us.

* **Comets**

  Occasional visitors to the Earth's neighbourhood. A comet grows a tail as it approaches the Sun, and this then shrinks as the comet travels away again over a period of months.

* **Stars**

  Hot balls of gas like the Sun, but at enormous distances. Over the course of a human lifetime they are fixed in relation to each other.

* **Nebulae**

  Faint misty patches that are actually vast glowing masses of gas among the stars.

* **Milky Way**

  A faint band of light circling the sky, the effect of looking through the breadth of the disc-shaped system of stars, gas and dust that is our Galaxy.

* **Galaxies**

  Misty patches, hard to tell apart from nebulae with the naked eye, but actually vast collections of stars, gas and dust beyond the Milky Way.

The Turning Sky_

Viewed from the turning Earth, everything in the sky seems to rise in the east and set in the west. The stars make the circuit in 23 hours, 56 minutes and 4.1 seconds; this is the time it takes the Earth to turn once. But because of the Earth's year-long movement around the Sun, the Sun seems to lag behind the stars, returning to the same place in the sky approximately four minutes later each day. So the time between one noon and the next is, on average, 24 hours. And in relation to the stars, the Sun drops back a little toward the west. It returns to its original place among the stars after a year.

The axis of the Earth is tilted in relation to the Earth's orbit around the Sun. In June the north pole is tilted towards the Sun, and there is summer in the northern hemisphere, and at the same time there is winter in the southern hemisphere. The angle of tilt does not change during the year, so when the Earth has moved round to its December position, the north pole is tilted away from the Sun, and it is winter in the northern hemisphere and summer in the southern hemisphere.

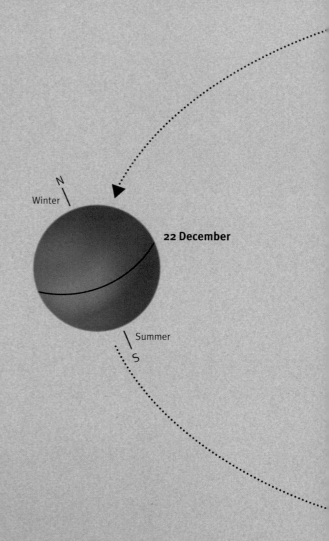

N

Winter

22 December

Summer

S

Something to Think About . . .

The Earth goes around the Sun in approximately 365¼ days. We put an extra day into the calendar every fourth year, when the year number is divisible by four (except for century years, when the year number must be divisible by 400). These 'leap' years keep the calendar in step with the seasons, which in turn keeps the calendar accurate to within a day every 3,226 years. A possible refinement would be to adjust 4000 AD, and all later years that are multiples of 4,000, so that they are *not* leap years. That would keep the calendar in step with the seasons to within a day for 16,667 years.

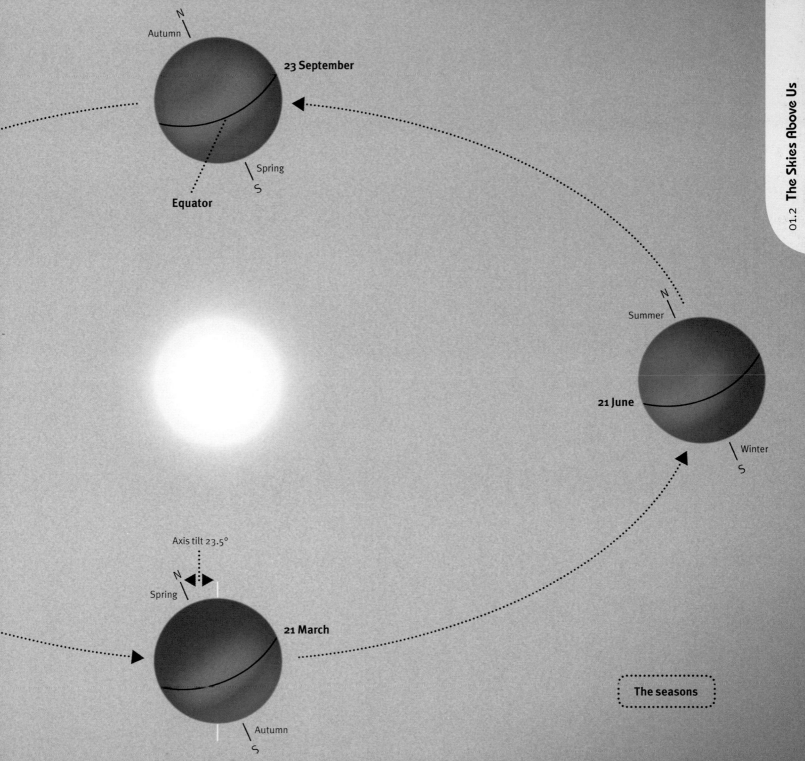

Autumn

N

23 September

Spring

S

Equator

Summer

N

21 June

Winter

S

Axis tilt 23.5°

N

Spring

21 March

Autumn

S

The seasons

17

Mapping the Sky_

Astronomers map the skies by imagining that celestial objects are fixed to the surface of the celestial sphere, a vast imaginary globe centred on the Earth. On the celestial sphere above the Earth's poles are the celestial north and south poles, while above the Earth's equator is the celestial equator.

The celestial sphere and the Sun, Moon and planets seem to revolve around the Earth from east to west – but this is just the effect of the Earth revolving from west to east every day.

The Sun also seems to move against the background of fixed stars in the course of a year. This is the effect of the Earth travelling around the Sun. The Sun's path is called the ecliptic. Because the Earth's equator is tilted in relation to its path around the Sun, the ecliptic is tilted in relation to the celestial equator.

In June the Sun is at its furthest north from the celestial equator, and the Earth's northern hemisphere receives more heat and light than the southern hemisphere – it is the northern summer. This position is called the solstice. The moment when the Sun reaches that position is also called the solstice. When the Sun is at the southern solstice in December, it is summer in the southern hemisphere.

Something to Think About . . .

Some people think that summer occurs because the Earth is actually closer to the Sun than it is in winter – forgetting that at the same time it is winter in the opposite hemisphere. Actually, the Earth is closest to the Sun early in January, in the middle of the northern winter, and furthest away in June, in the northern summer. This means that the northern seasons are slightly milder than they otherwise would be.

Armillary sphere

Astronomers of the past modelled the
imaginary reference lines in the sky with
the armillary sphere. The celestial equator,
the celestial poles, the ecliptic and other
points and lines were represented by an
assemblage of hoops and rings.

The Scale of the Universe_

The distances we encounter in our exploration of the Universe are overwhelmingly large. Astronomers use the speed of light to make distances comprehensible.

Light (and also radio waves and all other forms of electromagnetic radiation – see pages 58–59) is the fastest thing in the Universe. It travels at 300,000km (186,282 miles) per second, which means it can circle the Earth in one-seventh of a second. It travels to us from the Moon in 1.3 seconds and from the Sun in just over eight minutes. Signals from a spacecraft as far away as Neptune, the farthest planet, take over four hours to reach us.

But light travels for many years to reach us from the stars. The distance light travels in a year is 9.5 trillion km (5.9 trillion miles) and is called a light year. The closest star is 4.2 light years from us. (Astronomers also sometimes talk about light seconds, light minutes and light hours.) Stars in the Milky Way are tens of thousands of light years from us; other galaxies are millions of light years away. And at the farthest limits of observability are the intensely bright objects known as quasars. Their light takes over 11 billion years to reach us.

The Sun
8.3 light-minutes

The Moon
1.3 light-seconds

Earth

Something to Think About . . .

The Greek scientist Archimedes made an amazing estimate of the size of the Universe in the 3rd century BC. He calculated a diameter that in modern terms is equivalent to two light years. That's an infinitesimal fraction of the Universe we know today – yet still staggeringly huge by the standards of his day. Archimedes went on to estimate that the number of grains of sand needed to fill a universe this size was represented by a 1 followed by 63 zeros.

Quasar 4C 71.07
11 billion light years

Alpha Centauri system
4.2 light years

Andromeda Galaxy
2.5 million light years

A light second equals 300,000km (186,000 miles)
A light year equals 9.5 trillion km (5.9 trillion miles)

01.5 Building Towards the Sky_

Ancient peoples recorded their knowledge of the changing skies in written documents and oral myths – and sometimes also in impressive structures. The Great Pyramid at Giza in Egypt was built around 2560 BC. Astronomical skill was needed to align its sides precisely north–south and east–west. In the 13th century BC the Egyptian pharaoh Ramesses I built the temples at Abu Simbel. Sunlight enters the tomb twice a year: on 21 October, the pharaoh's birthday, and on 21 February, the date of his coronation. It falls on statues of the pharaoh flanked by two gods.

Around AD 1000, Pueblo Indians set three slabs of rock upright on a rocky outcrop called Fajada Butte, in Chaco Canyon, New Mexico. The Sun shining between the slabs casts 'sun daggers' – bars of sunlight – on to a spiral petroglyph, a symbol engraved in a rock face. Particular positions of the daggers on the pattern indicate midsummer, midwinter and the equinoxes.

Around 300 BC a large mound tomb was built at Newgrange in the east of the Republic of Ireland. The entrance passage is precisely aligned to the direction of sunrise at the winter solstice, and light can enter only briefly for a few minutes on the few days around that date.

Perhaps the best-known ancient building with an astronomical purpose is Stonehenge in the south of England. It is still impressive today, even though it is a mere remnant of the great complex of timber and stone circles that once stood there. Earth and timber constructions date from 3000 BC, but stones were erected from about 2600 BC. The largest stones weigh up to 50 tonnes. There is a long-held idea that some were transported 250km (155 miles) from Wales, but this is highly controversial.

Stonehenge doubtless served many social and religious purposes. Most of these remain mysterious, but some rituals seem to have been connected to the midsummer solstice. The structure's main entrance was towards the northeast, oriented towards the midsummer sunrise. On midsummer's day, from inside Stonehenge, it is possible to see the Sun rise over the Heelstone that lies outside the stone circles.

Something to Think About . . .

Hundreds of alignments of various Stonehenge features with important astronomical directions have been found, and inevitably some have said that alien intelligences must have given a little help to the builders. But actually, bogus patterns are not hard to find just by chance when you study a large collection of objects. In 2010 a mathematician, Matt Parker, looked at the positioning of 800 significant cultural sites scattered across Britain, and found numerous alignments and patterns. The sites were Woolworth stores.

Pictures in the Sky_

01.6

To most of us a constellation represents an imaginary figure in the sky based on some fancied resemblance – usually pretty far-fetched – to a human being, animal or object, real or mythological. But in astronomy it technically means one of 88 internationally defined areas into which the sky is divided. Of these, 48 correspond to constellations recorded in the 2nd century AD by the Greek astronomer Claudius Ptolemy, and many of those had come to him from Mesopotamian astronomers dating from before 1000 BC. Later astronomers altered Ptolemy's constellations.

When Europeans sailed south of the equator, they observed the southernmost sky for the first time, and devised new constellations, including modern objects such as the Clock, Microscope and Telescope.

Each constellation has a Latin name. The brightest stars in a constellation each have a traditional name and a scientific name. The scientific name is made up from the letters of the Greek alphabet – Alpha, Beta, Gamma and so on – combined with a form of the constellation's Latin name. So the brightest star in Leo has the traditional name Regulus and the scientific name Alpha Leonis ('Leonis' means 'of the lion'). The constellations are shown on the star maps on pages 218–19.

Leo

Ursa Major

Leo Minor

Something to Think About . . .

Aboriginal Australians populate their sky with star figures such as a canoe, which is part of our constellation Orion. But they also form figures from the dark areas in the band of light that is the Milky Way. One is called the Emu.

Sky figures

The constellations were traditionally represented in star atlases by beautiful and elaborate pictures. In this functional age, stick figures are usually used to link the main stars of the constellation. (The constellations here are not shown in their true positions relative to each other.)

Pegasus

Orion

Gemini

Canis Major

Cancer

01.7 The Northern Sky_

During the course of a year someone in the northern hemisphere can see all of the northern half of the sky and some of the southern half. If they live at 40° north – the latitude of southern Europe or the middle of the USA – they can see as far as 50° south of the celestial equator. The stars within 40° of the celestial north pole never set for this observer – they are called circumpolar stars. The most conspicuous of these constellations are the Great Bear and Cassiopeia. The Pole Star – also called the North Star or Polaris – lies close to the pole, at the tip of the tail of the Little Bear.

Constellations further south change through the seasons. In summer, three bright stars called the Summer Triangle, belonging to different constellations, appear high in the sky: Altair, in Aquila, the Eagle; Deneb, in Cygnus, the Swan; and Vega, in Lyra, the Lyre. In winter Orion, the Hunter, dominates the sky.

The Great Bear

The familiar stars of the Plough or Big Dipper are an asterism – a distinctive group of stars not officially classed as a constellation. The Plough is just part of its host constellation, the Great Bear. The Plough revolves anticlockwise around the celestial north pole every day. The two stars at its end, called the Pointers, point roughly towards the pole.

Something to Think About . . .

Before reliable clocks and watches were invented, people would tell the time at night using the position of the stars, just as they used the position of the Sun during the day. A 'nocturnal' was a pocket device on which you would twist the dial to set the date, sight the Pole Star through a hole in the centre, and line up a pointer with selected stars – usually the stars in the Plough called the Pointers. Then you would be able to read off the time from the pointer. This 'star-dial' was more accurate than a sundial, because the stars keep better time than the Sun.

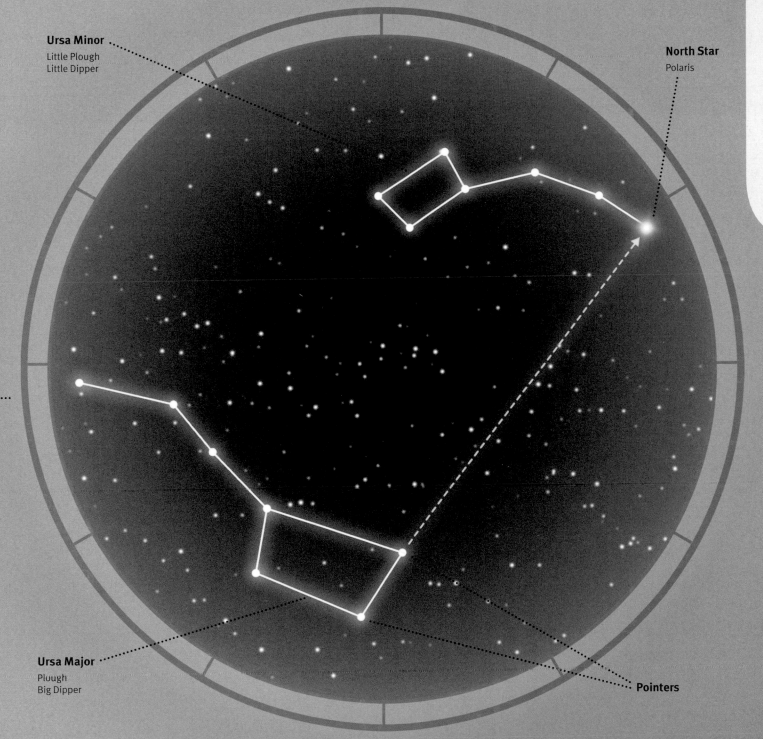

Ursa Minor
Little Plough
Little Dipper

North Star
Polaris

Ursa Major
Plough
Big Dipper

Pointers

The Southern Sky_

Seen from the southern hemisphere, it is the celestial south pole that the sky appears to revolve around. There is no bright star near this point, but travellers are guided by Crux, the Southern Cross, a small bright constellation. For an observer at 40° south, the latitude of southern Australia and southern South America, all the stars within 40° of the pole are circumpolar. These constellations include Carina, the Keel, and part of Centaurus, the Centaur.

Looking towards the celestial equator, observers at 40° south can see the sky as far as 50° north, so they see the same ever-changing mid-sky constellations as someone at the mid-latitudes of the northern hemisphere. But they see the stars rising on their right and setting on their left during the night – the reverse view to those in the north.

Something to Think About . . .

When you navigate using the stars in the southern hemisphere, you can be misled by the False Cross. This is not a constellation – a defined astronomical area within the night sky – but an asterism, simply a pattern of stars, consisting of four stars from the constellations of Carina, the Keel, and Vela, the Sails. The grouping is fainter and more spread out than the Southern Cross, and has four main stars, rather than five.

Gamma

Delta

· · · · · **Crux**

Beta

Epsilon

The Southern Cross

Crux, the Southern Cross, is the smallest constellation, but one of the brightest. Follow the line of the cross's longer arm to 4.5 times its length and you are close to the pole. Australians think of it as 'their' constellation, and it appears on their national flag.

Alpha

The Zodiac: Circle of Life_

Even astrology-haters are familiar with the names of the 12 constellations of the zodiac. The word 'zodiac' ultimately derives from the Greek for 'living thing', because all but one of the figures represent real or mythological creatures or people. The zodiac consists of 12 constellations through which the Sun appears to pass in a year.

Astrologers divide the Sun's path into 12 divisions, called signs, corresponding to the constellations. But the signs are of equal size, whereas the sizes of the constellations are very different. Astrologers claim that a person's character and fate are influenced by the sign that the Sun is in at the moment of their birth. Curiously, the astrologers' signs have got out of step with the sky. A 'wobble' in the Earth's axis, called precession, means that the Sun enters each sign about a month later every 2,150 years. The influence of the sign of Cancer, for example, is said to begin about 23 June, but nowadays the Sun enters the constellation of Cancer about 21 July.

Something to Think About . . .

A 13th constellation, Ophiuchus, the Serpent-bearer, intrudes into the zodiac between Scorpio (the astrological name for Scorpius, the Scorpion) and Sagittarius (the Archer). Some astrologers use a zodiac of 13 signs. And there's even a 14-sign zodiac, incorporating Cetus, the Whale, whose tail lies between Pisces (the Fishes) and Aries (the Ram). But no matter how much astrologers update their ideas, scientists will never take them seriously.

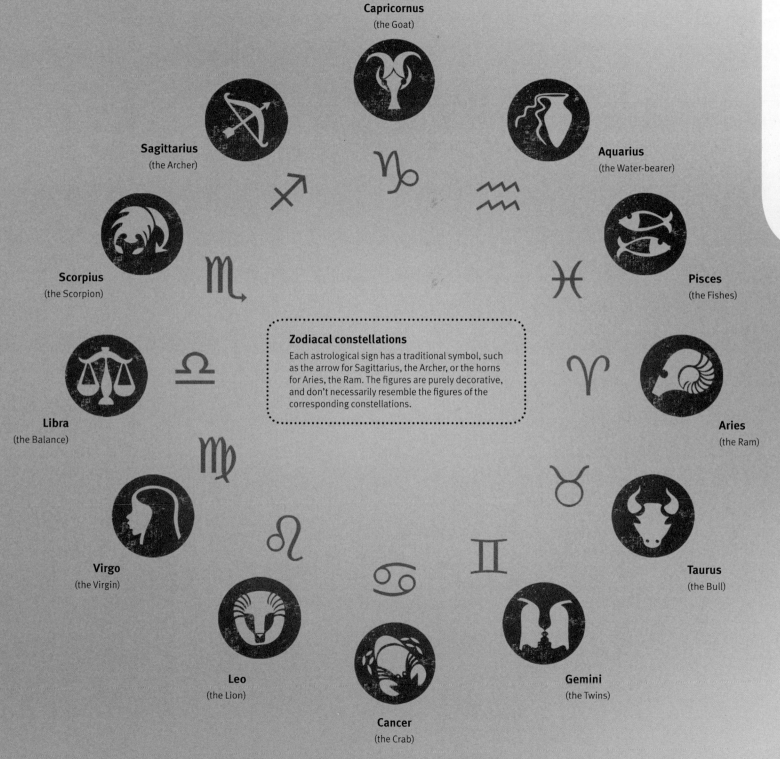

Capricornus
(the Goat)

Sagittarius
(the Archer)

Aquarius
(the Water-bearer)

Scorpius
(the Scorpion)

Pisces
(the Fishes)

Zodiacal constellations
Each astrological sign has a traditional symbol, such as the arrow for Sagittarius, the Archer, or the horns for Aries, the Ram. The figures are purely decorative, and don't necessarily resemble the figures of the corresponding constellations.

Libra
(the Balance)

Aries
(the Ram)

Virgo
(the Virgin)

Taurus
(the Bull)

Leo
(the Lion)

Gemini
(the Twins)

Cancer
(the Crab)

01.10 Man-made Skies_

Human beings have tried to grasp the nature of the skies in maps and globes. But the pinnacle of the human mind's conquest of the sky is to create a working model of its intricate movements. A medieval instrument called an astrolabe pictured the turning of the sky, with especially beautiful examples being made by Arab craftsmen.

The development of clockmaking from the 17th century made mechanical models of the Solar System possible, trading on the new understanding that the Earth moves around the Sun. Elaborate hand-cranked machines showed the Earth, Moon, planets and newly discovered satellites of other planets, all rotating around a central Sun. The true relative sizes of the orbits could not be squeezed into a table-top model, but relative speeds could be shown – Mars, for example, taking roughly twice as long as the Earth to revolve around the Sun.

In the 20th century electronics could create the illusion of sitting under the open sky in the comfort of a planetarium. A modern planetarium doesn't merely display images of the naked-eye stars as seen from Earth, but can take the audience on a 'virtual-reality' flight through the stars and beyond our galaxy.

Mars

The Moon

The Sun

Something to Think About ...

A badly corroded, bronze, disc-shaped mechanism was raised from an underwater shipwreck off the Greek island of Antikythera in 1902, where it had lain for over 2,000 years. After decades of research, archaeologists realized it was in fact an astonishingly sophisticated mechanical computer. By entering a date into the machine, the positions of the Sun, Moon and possibly the five naked-eye planets were shown, and eclipses could be predicted.

Saturn

Uranus

Neptune

A planetarium

Here an audience can watch the movements of a simulated sky. Traditionally images were projected by a huge, rotating, insect-like projector. More modern planetaria use inconspicuous digital projectors.

Through the Telescope_

When you look up at the night sky through a pair of binoculars or a telescope – even a small one – the view is transformed.

··÷ Thousands more stars are revealed. Compare, for instance, thenumber of stars you can see within the large square of the Plough compared with the few you can see with the naked eye.

··÷ As stars look brighter, your eyes are able to register their colours. But the stars remain points of light.

··÷ The Milky Way is no longer a band of light but a field of thousands of faint stars.

··÷ Clusters of stars become richer. Some stars are seen to be multiple, consisting of two, three or even more stars.

··÷ The Moon becomes a world of craters, mountains and plains.

··÷ The planets are transformed from points of light to worlds showing discs.

··÷ Nebulae and galaxies become brighter, larger and more numerous – though only powerful telescopes can really show their structure.

A telescope or pair of binoculars, used as far from city lights as possible, will vastly increase the number and clarity of the objects you can see in the night sky.

Warning
Even glimpsing the Sun briefly through a telescope or binoculars can damage your eyesight permanently. Turn instruments to the sky only after the Sun has set or well before it rises. For advice on using a telescope to observe the Sun, see page 102.

Star clusters look impressive through the telescope. They are often given descriptive names such as the Jewel Box and the Beehive.

Something to Think About . . .

A telescope designed for purely astronomical use will produce an upside-down image. This does not matter at all; the extra lenses needed to turn the image the right way up would very slightly reduce the quality of the image. Printed maps of the Moon traditionally showed south at the top of the image to match what observers would see through their telescopes.

01.12 Sky-watching With the Naked Eye_

It is possible to find fascinating things to look at in the sky every night of the year without even picking up a telescope or a pair of binoculars.

Learning the constellations

···> Get hold of a compact astronomical observing book that you can consult by torchlight while you are stargazing.

···> A planisphere is a handy guide to the changes that take place in the night sky over the course of a year. A clear window representing the visible sky rotates over a star map. Set the window to your date and time and you can see which stars are above the horizon at that moment.

···> If you have a smartphone, you could get a stargazing app for it. Some astronomy apps – for phones that have built-in magnetic direction sensors – will identify the objects in an area of sky when you hold the phone up towards that part of the sky.

···> Learn the main 'signpost' stars and constellations near the celestial pole, which you'll see all year round. Learn the stars and constellations nearer the equator as they appear season by season.

Stargazing in company is fun and educational. It can also encourage you to travel further afield in search of darker skies.

Something to Think About . . .

Get as far away as you can from city lights to view the stars. John E. Bortle's Dark Sky Scale rates sky darkness from 1 (the darkest) to 9 (the most light-polluted). You can see a Class 1 sky only from mid-ocean, mountaintops or wilderness areas. Here the brightest parts of the Milky Way cast visible shadows. In a Class 6 sky ('bright suburban'), the Milky Way is barely visible.

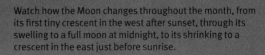

Watch how the Moon changes throughout the month, from its first tiny crescent in the west after sunset, through its swelling to a full moon at midnight, to its shrinking to a crescent in the east just before sunrise.

Events in the sky

* Check the regular astronomy features in your newspaper and on websites such as astronomynow. com, skyandtelescope.com, and astronomy.com to see which planets and constellations will be conspicuous during the next month.

* Showers of meteors occur at regular times during the year – your news sources will alert you to them.

Sky-watching With Instruments_

If you already use binoculars for sports or bird-watching, you can turn them on the sky tonight to give a breathtaking enhancement of your view of the Universe. The most serious amateurs use telescopes, but before spending money on a new instrument, make sure you do your research carefully so you get the best one for your needs.

Binoculars and telescopes are assessed by their magnifying power and the size of their objective (main) lens or mirror. The larger the magnification, the larger the images of the Moon, the planets and other extended objects, such as star clusters, nebulae and comets, that you will get. But a star remains a point of light, however great the magnification. The larger the objective, the greater the light-gathering power of the instrument, and the brighter the image.

You might think you want as much magnification as possible, but it can be harder to find your way around an area of sky if you can only see a small part of it. Large magnification also makes an image dimmer, and so calls for larger light-gathering power. It also makes it more necessary to support the instrument to prevent image shake. Binoculars with image stabilization are good for astronomy, but expensive.

A telescope may be a reflector or a refractor (see pages 52–53). Some popular amateur telescopes combine a reflector and a refractor but more advanced astronomical telescopes tend to be reflectors.

Something to Think About . . .

Inexpensive telescopes are often misleadingly advertized as having large objective sizes. Behind the low-quality objective is an iris, a ring that cuts down the effective breadth of the lens to the central part. This removes some of the defects in the image – but at the cost of giving a fainter image than the advertized size suggests.

As you become more serious about astronomy you will acquire more and more equipment. This can include:

* A tripod.
* A range of eyepieces, giving different magnifications.
* A camera attached to the telescope.
* A motor drive to move the telescope to keep track of the object as it moves.
* Computerized aiming of the telescope. The user keys in the name of the object or position to view, and either a screen shows which way to move the telescope manually, or a motor does it automatically.

Ordinary digital cameras or special astronomical cameras can be connected to telescopes. Images can be built up over hours throughout the night.

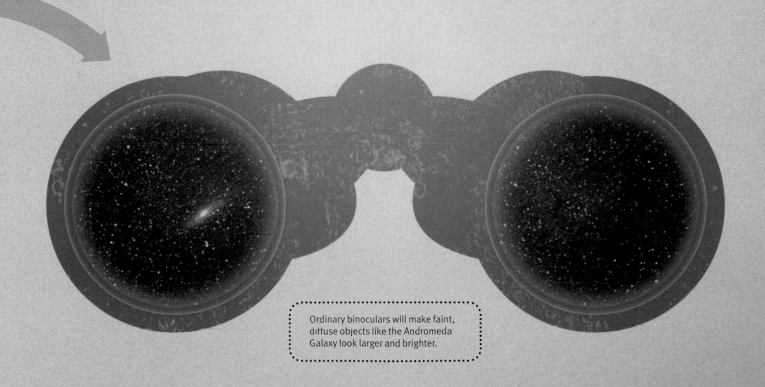

Ordinary binoculars will make faint, diffuse objects like the Andromeda Galaxy look larger and brighter.

Chapter 02.0

Breakthroughs_

Wheels Within Wheels_

Very early on, some Greek thinkers conjectured that the Earth rotates and moves through space. In the 5th century BC, the philosopher Philolaus taught that not only the Earth, but the Sun, Moon and planets, revolve around a great central fire, hidden from us because our side of the Earth is always turned away from it. In the 3rd century BC the Greek astronomer and mathematician Aristarchus of Samos, taught that the Sun is central and fixed, and that the Earth and the other planets revolve around it. He was accused of impiety for setting the Earth in motion.

In fact Aristotle had already developed more influential doctrines in the 4th century BC. In his cosmology, the skies consisted of dozens of interconnected spheres, whose regular circular movements gave a complicated motion to the planets that were fixed to certain of the spheres.

Aristotle's picture of the Universe was made mathematical by Ptolemy, the greatest of the ancient astronomers, who lived in the 2nd century AD. Ptolemy tried to combine simple circular motions to produce the complicated movements of the planets. Each planet revolved, not directly around the Earth, but in a circular orbit called an epicycle. The centre of the epicycle in turn revolved around the Earth. With enough tweaking, Ptolemy's system could predict the planets' movements far better than any of its predecessors, and it prevailed in astronomy until the 16th century.

Something to Think About . . .

To get the planet's motions right, Ptolemy had to make each planet's circular main orbit eccentric – that is, off-centre from the Earth, by a different amount for each planet. And worse still, the planets had to move around their circles at varying speeds, which outraged some thinkers' ideas of the perfection that heavenly bodies should possess.

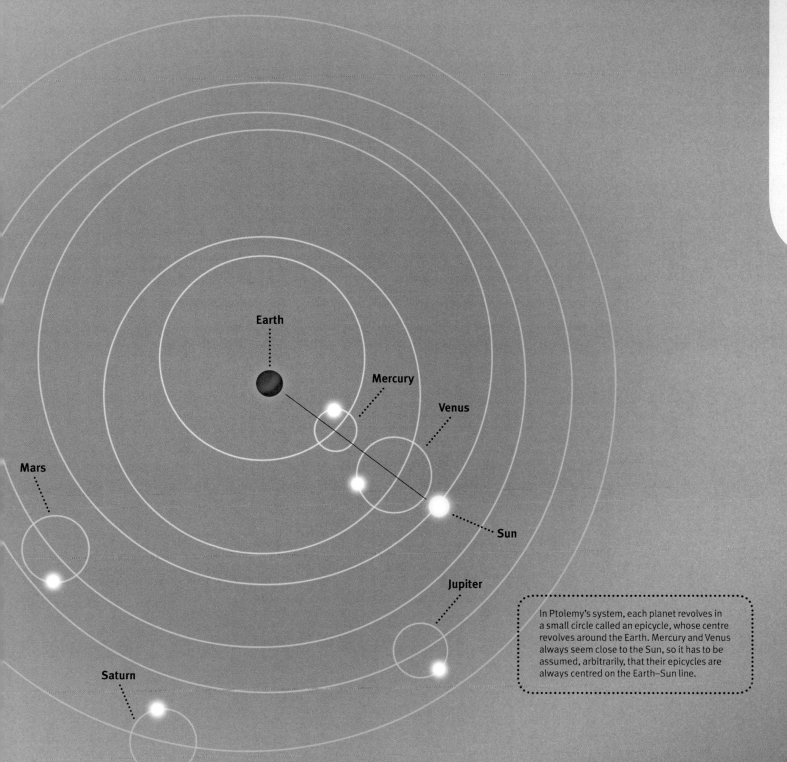

Earth

Mercury

Venus

Mars

Sun

Jupiter

Saturn

In Ptolemy's system, each planet revolves in a small circle called an epicycle, whose centre revolves around the Earth. Mercury and Venus always seem close to the Sun, so it has to be assumed, arbitrarily, that their epicycles are always centred on the Earth–Sun line.

The Earth Moves_

In the 15th century, after Ptolemy's ideas had held sway for 1,300 years, a Polish-German churchman, Nicolaus Copernicus, transformed astronomy and launched the Scientific Revolution.

Copernicus didn't object to Ptolemy's use of cycles and epicycles. But he strongly objected to Ptolemy's use of non-uniform speeds for the planets. He claimed the whole system would be much simpler if the Sun were taken as fixed at the centre of the Universe, or rather, near it. To make things work he had to shift the Sun off-centre. Copernicus used only perfect circles in his description of the planets' motion, and he used even more epicycles than Ptolemy.

Although the details of his system were at least as complicated as Ptolemy's, Copernicus introduced important simplifications into the overall picture of the Solar System.

⋯⇢ The orbits of Mercury and Venus are inside the Earth's orbit. This explains why they always seem close to the Sun.

⋯⇢ The outer planets seem to slow down and go backwards at times. (This is called retrograde motion.) It occurs only when the planet is opposite to the Sun in the sky, and Ptolemy didn't have a convincing explanation of why this should be so. The Copernican theory explains this as due to the Earth 'overtaking' the slower-moving outer planets.

Copernicus described his ideas in lectures, but he withheld publication of his book on the system until the year of his death. His ideas were freely discussed until, in the following century, the Church began to restrict the teaching of the Sun-centred Universe.

Something to Think About . . .

In the Copernican system it is very easy to work out the relative sizes of the planets' orbits by observing retrograde motion. Thus Copernicus found that Jupiter is five times as far from the Sun as the Earth is. But no one could yet calculate just how far that actually was.

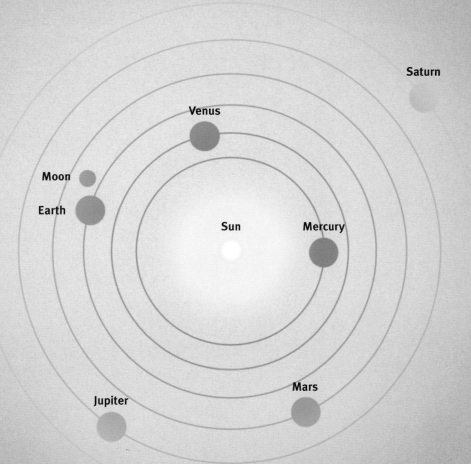

Saturn

Venus

Moon

Earth

Sun   Mercury

Jupiter

Mars

In the Copernican system, the Sun is lord of the
Solar System. All the planets revolve around it;
only the Moon revolves around the Earth.

45

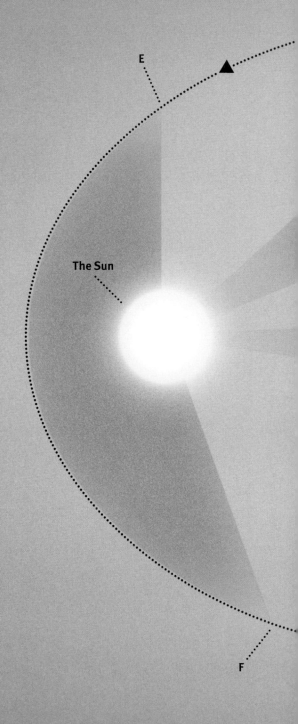

02.3 Laws of the Planets_

In 1600 a brilliant German astronomer, Johannes Kepler, began working as an assistant to Tycho Brahe, the last and greatest naked-eye mapper of the sky. When Brahe died, Kepler took over his highly accurate observations and, after epic labours, forged three great laws of planetary movement.

···❯ Kepler's First Law broke away from astronomers' millennia-long obsession with perfect circles. It states that every planet moves in an ellipse – a flattened circle. The Sun does not even lie at the centre of ellipse, but off to one side, lopsidedly.

···❯ The Second Law states that a line joining a planet and the Sun sweeps out equal areas during equal intervals of time. This means that a planet must move faster when it is closer to the Sun (see illustration).

···❯ The Third Law relates the time it takes a planet to circle the Sun – its 'year' – to its distance from the Sun.

Kepler's picture of the planets' movements was radically new, simple and elegant. Now it just needed some way of explaining why the planets should behave like this.

Something to Think About . . .

Kepler wrote possibly the first science-fiction story. *The Dream* tells how an astronomer is helped on a journey to the Moon by his mother, a witch. The story created problems for Kepler's own mother and she was put on trial for witchcraft. Kepler defended her and got her released.

E

The Sun

F

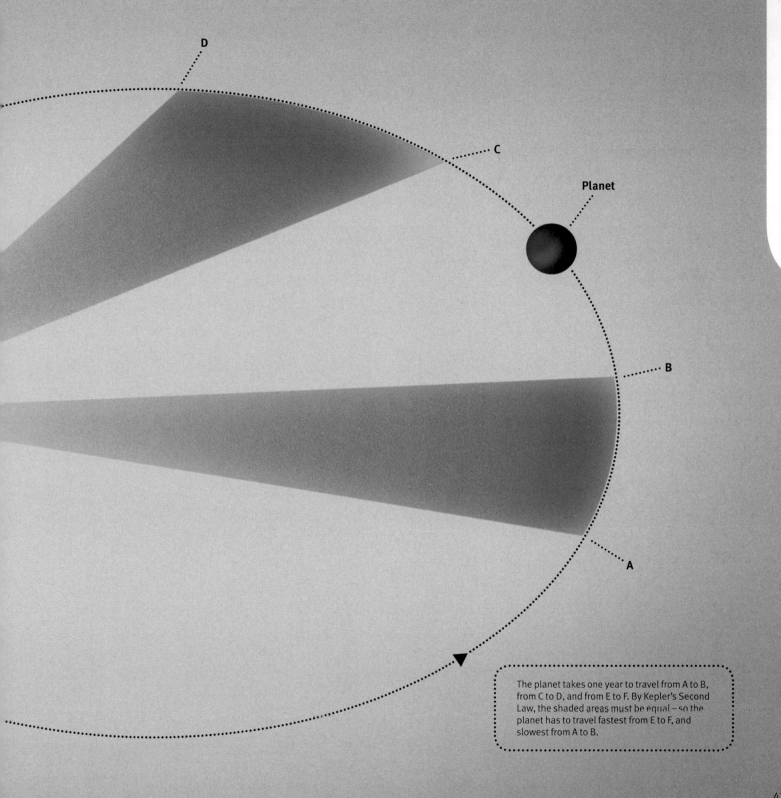

D

C

Planet

B

A

The planet takes one year to travel from A to B, from C to D, and from E to F. By Kepler's Second Law, the shaded areas must be equal – so the planet has to travel fastest from E to F, and slowest from A to B.

Telescopes on the Heavens_

In 1609, a cantankerous professor of physics in Florence called Galileo Galilei heard of a new device invented by a Dutch spectacle-maker. Called the 'telescope', it consisted of lenses mounted in a tube, and could reputedly make distant objects seem closer. From this description alone, Galileo made his own telescope and soon turned it to the sky. He gathered devastating ammunition in his battle against the ancient physics and astronomy that still prevailed.

Here are some of the things Galileo discovered:

···> The sky is filled with scores of thousands of stars too faint to be seen with the naked eye. Why revere the ancients when they had known nothing of all this?

···> The planet Venus shows phases like the Moon, proving that it revolves around the Sun – not between the Earth and the Sun, as Ptolemy had thought.

···> There are spots on the Sun, which traditionally was supposed to be perfect.

···> Four satellites circle Jupiter – a possibility not conceived of in the traditional astronomy.

Something to Think About . . .

Galileo observed the planet Neptune on two occasions. But he had no way of knowing it was not just a faint star. It was discovered and recognized as a planet in 1846.

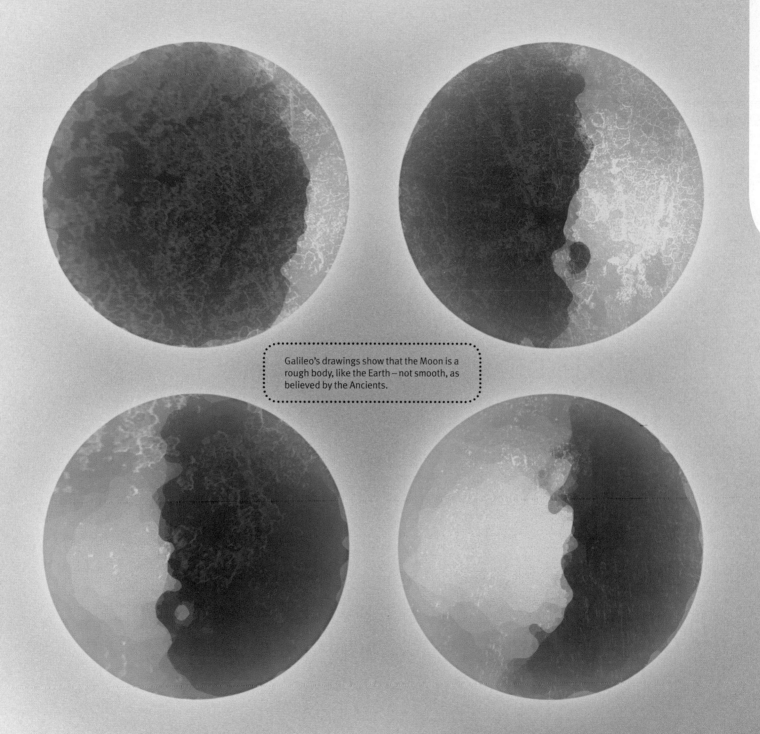

Galileo's drawings show that the Moon is a rough body, like the Earth – not smooth, as believed by the Ancients.

Universal Attraction_

The discoveries of Copernicus, Kepler and Galileo all made sense after Isaac Newton published his Law of Universal Gravitation in 1684.

Newton made an astounding claim: every piece of matter in the Universe attracts every other. So not only does the Earth pull on you, but so do the Moon and Sun and planets, though the effects of these distant bodies aren't noticeable.

Also, each body in the Solar System pulls on all the others. This explains why Jupiter manages to keep its satellites revolving around it, as Galileo had discovered. The Sun and the Earth both pull on each other. The reason the Earth goes around the Sun, but the Sun doesn't go around the Earth, is that the Sun has much greater mass.

An object's mass is, roughly, the amount of matter in the object. It shows itself in two ways: as how strongly the object pulls on another one gravitationally; and second how much it resists being moved. Because of its huge mass the Sun both has a strong gravitational pull and resists being pulled. Nevertheless, it does wobble a bit because of the pull of all the planets, including the Earth.

Something to Think About . . .

In 1798 the eccentric English nobleman Henry Cavendish 'weighed the world' by measuring the attraction between a pair of large lead spheres and a pair of small ones swinging on a suspended rod. He compared the feeble effect with the strength of the Earth's gravity and found the mass of the Earth within 1 per cent of the modern value (which is 5.9742×10^{24} kg).

$$F_g = G\,\frac{m_1 m_2}{r^2}$$

$$F_1 = F_2 = G\,\frac{m_1 \times m_2}{r^2}$$

Newton's law of gravitation says that an object's pull on a second object:

* increases with the mass m_1
* increases with the mass m_2
* decreases with the square of the distance, r^2 (if you double r the force decreases to a quarter; if you triple r the force decreases to a ninth, and so on)

G is a number called the universal constant of gravitation.

The attraction between the two objects will tend to pull them together. But if they have some sideways movement, they'll probably end up orbiting around each other – as stars and planets and other objects in space do.

Key
F = the force between the masses
G = the gravitational constant
m_1 = the first mass
m_2 = the second mass
r = the distance between the masses

How Telescopes Work_

A telescope that uses lenses is called a refracting telescope, or refractor, because the lenses refract or bend light. The main lens is called the objective and additional lenses are used in the eyepiece. Galileo's telescopes consisted of just two lenses and produced low-quality images, but gave him a magnifying power of up to 20 times.

Early refractors suffered badly from image defects. One problem was that simple lenses form colour fringes. This was eventually overcome by combining lenses made from different sorts of glass. But another approach was invented by Isaac Newton. Mirrors don't form colour fringes, so Newton used a concave mirror instead of the objective lens to form an image, and then viewed this through an eyepiece.

Refractor

The simplest refracting telescopes have two lenses. One convex lens brings the light from an object towards a focus, a second lens forms the eyepiece. Galileo's telescopes had a single concave lens for an eyepiece, but an eyepiece can be a complex construction of several lenses.

Something to Think About . . .

The largest telescopes are those used by astronomers. They are always reflectors, because a mirror can be supported over its whole area from behind. A lens of equal diameter could only be supported at its edges and would be distorted by its own weight.

Reflector

The simplest reflecting telescope uses a concave mirror to bring light from the object towards a focus. A flat mirror diverts the light out of the telescope tube, and an eyepiece is mounted at the side.

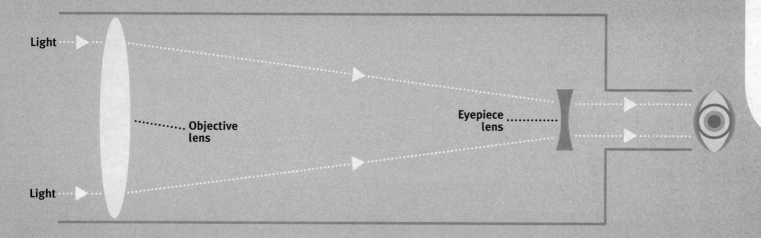

Light ▷

Light ▷

.... Objective
lens

Eyepiece ....
lens

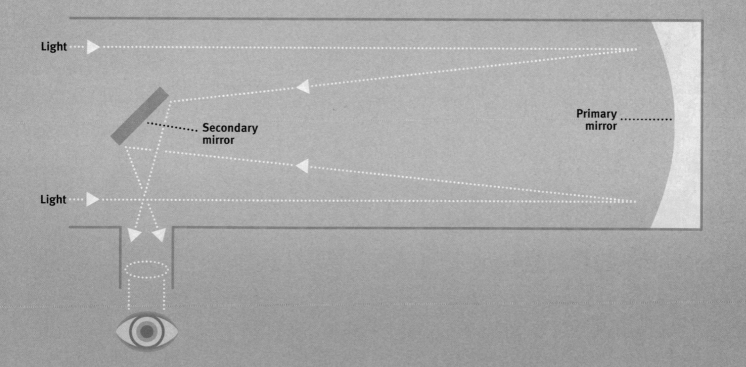

Light ▷

Light ▷

.... Secondary
mirror

Primary ....
mirror

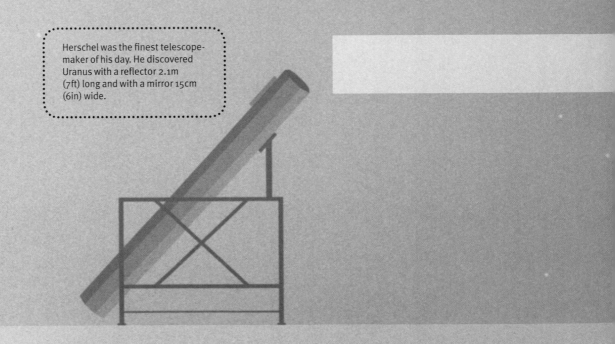

Uranus: the First New Planet_

In 1781 the telescope, which had revealed new stars and new satellites, produced another triumph. The discovery of a new planet was announced.

William Herschel was born Friedrich Wilhelm Herschel in 1738 in Hanover, but lived in England from the age of 19, and became perhaps the greatest astronomical observer of all time. He noticed that a certain 'star' was a fuzzy disc in the telescope. Herschel reported it as a comet, but within a couple of years other astronomers, and finally Herschel himself, decided that it must be a new planet. Its name, after much indecision, was finally settled upon: Uranus.

Curiously, the new planet had always been visible to the naked eye, but there was nothing to distinguish it from the hosts of faint stars. And since it takes a human lifetime for Uranus to creep once around the sky, its motion among the stars was never noticed.

Herschel was the finest telescope-maker of his day. He discovered Uranus with a reflector 2.1m (7ft) long and with a mirror 15cm (6in) wide.

Something to Think About...

Herschel first suggested calling the new planet by the Latin name *Sidum Georgium*, the Star of George, after King George III. Coincidentally, Herschel ended up with a knighthood, a pension, and the title of 'King's Astronomer'. Astronomers of other nations were less enamoured by the name and eventually settled on Uranus.

To Herschel, Uranus appeared as just a tiny blue featureless disc.

In 1979 astronomers using an airborne observatory discovered rings around Uranus. These poor relations of Saturn's rings were so faint they could only be seen as they darkened stars beyond. Yet Herschel had described a faint ring around Uranus in 1797. Was he 180 years ahead of his time?

Asteroids: Vermin of the Skies_

At the end of the 18th century, there seemed to be a large gap in the Solar System between the orbits of Mars and Jupiter. An international group of 25 astronomers – the 'Celestial Police' – was organized to sweep the skies for some faint, undiscovered planet. But, as in so many crime thrillers, the 'police' were beaten to their quarry by a rival detective. In 1801 an Italian priest, Giuseppe Piazzi, discovered a tiny body that he called Ceres, after the Roman goddess of grain. Soon the Celestial Police discovered more and more such small bodies, and now millions are detected by automated systems each year.

Something to Think About . . .

Later the camera joined the hunt for asteroids, making its first find in 1891. Photographs of the sky close to the ecliptic (the band in which the planets and most asteroids move) were soon cluttered with so many asteroid trails that they were dubbed the 'vermin of the skies'.

Mercury
TBR 0.4
AU 0.39

Venus
TBR 0.7
AU 0.72

Earth
TBR 1.0
AU 1.00

Mars
TBR 1.6
AU 1.52

Ceres
TBR 2.8
AU 2.77

Titius-Bode Rule

Two German astronomers, Johann Titius (in 1766) and Johann Bode (in 1772), described a simple rule, the Titus-Bode Rule (TBR), that described the mean distances from the Sun of the planets (then only known as far as Saturn).
A series of numbers is calculated as follows:

* Start with the numbers 0 and 3

* Double each number after that to give: 0, 3, 6, 12, 24, 48, 96, 192, 384, 768

* Add 4 to each number to give: 4, 7, 10, 16, 28, 52, 100, 196, 388, 772

* Divide by 10.

The results quite closely matched the distances in terms of the Astronomical Unit (AU), with a gap between Mars and Jupiter. Astronomers became hugely interested in the rule when Uranus was discovered in 1781 at just the right distance, and still more so with the discovery of the first asteroid, Ceres, in 1801. But the rule flunked its next test when Neptune was discovered in 1846 at a distance of 30 AU, where Titus-Bode predicted approximately 39. And Pluto – discovered in 1930 and long regarded as a planet – failed hopelessly. No astronomer takes the Titus-Bode 'Law' seriously nowadays.

One Astronomical Unit (AU) equals the distance between the Earth and the Sun – about 149,597,871km (92,955,807 miles).

Titius-Bode prediction

Actual distance

Jupiter
TBR 5.2
AU 5.20

Saturn
TBR 10.0
AU 9.54

Uranus
TBR 19.6
AU 19.2

Neptune
TBR 38.8
AU 30.06

Pluto
TBR 77.2
AU 39.44

02.9 Invisible Light_

In a rainbow, nature separates sunlight into its component colours. Nineteenth-century scientists did the same by passing a ray of sunlight through a triangular glass block called a prism. The emerging ray fanned out into a band of colours called a spectrum. In 1801, William Herschel, the astronomer who discovered Uranus, was studying the warming effect of different colours of light. He was amazed to find the strongest warming effect occurred when he placed his thermometer on unseen rays beyond the red end of the spectrum. This invisible light was later called infrared ('below red') light.

Light consists of waves. The shortest wavelength of visible light is at the violet end of the spectrum (380 millionths of a millimetre). The longest is that of red light (about 700 millionths of a millimetre). Infrared light ranges in wavelength from this to about a millimetre (a purely arbitrary cut-off point). Still longer wavelength radiations were to be discovered: microwaves and radio waves.

In 1801 invisible rays were also being discovered beyond the violet end of the spectrum. A German scientist, Johann Ritter, discovered that unseen rays here could darken certain chemicals. Waiting to be discovered beyond ultraviolet rays lay X-rays and the even more penetrating gamma rays (emitted in some forms of radioactivity).

Something to Think About . . .

Visible light and the invisible radiations beyond the visible spectrum are now described as electromagnetic (EM) radiations, because they have electrical and magnetic effects, and can be produced by electrical and magnetic means. In astronomy, every wavelength band of the EM spectrum has its own detectors, which reveal new facets of the Universe.

White light consists of a mixture of different wavelengths (colours), which can be separated by a prism. It also contains invisible wavelengths – of ultraviolet and infrared light.

02.10 Measuring the Universe_

In 1908, Harvard College Observatory, Massachusetts, was well equipped with computers. It wasn't magically ahead of its time: 'computer' was the usual term then for a human being of lowly status who was employed to perform calculations. Henrietta Leavitt, a computer of 15 years' standing, in that year published measurements made on photographs of 1,777 variable stars. These are stars that change in brightness over time. Some variables brighten and dim in regular cycles with a definite period (maximum-to-maximum time). Leavitt mentioned that she had noticed that brighter stars had longer periods. Four years later she confirmed the relationship. Such stars are called Cepheid variables, or simply 'Cepheids', after one such star, Delta Cephei.

In 1913 astronomers measured the distance of a particular Cepheid. This meant that from its apparent brightness – how bright it looks to us – they could calculate its true brightness. Then, from Leavitt's brightness–period relationship, they were able to work out the true brightness of *any* Cepheid. And knowing its true brightness, by comparison with its apparent brightness, they could work out its distance.

When astronomers started using this new method of distance measurement, they were to find that the Universe was far larger than they had thought.

Something to Think About . . .

Astronomers had an unfortunate time with Cepheids at first. No one realized that there are at least two types of Cepheid, with different relationships between brightness and period, which meant that all the early calculations of distance were half what they should have been. It wasn't until the 1940s that the problem was resolved.

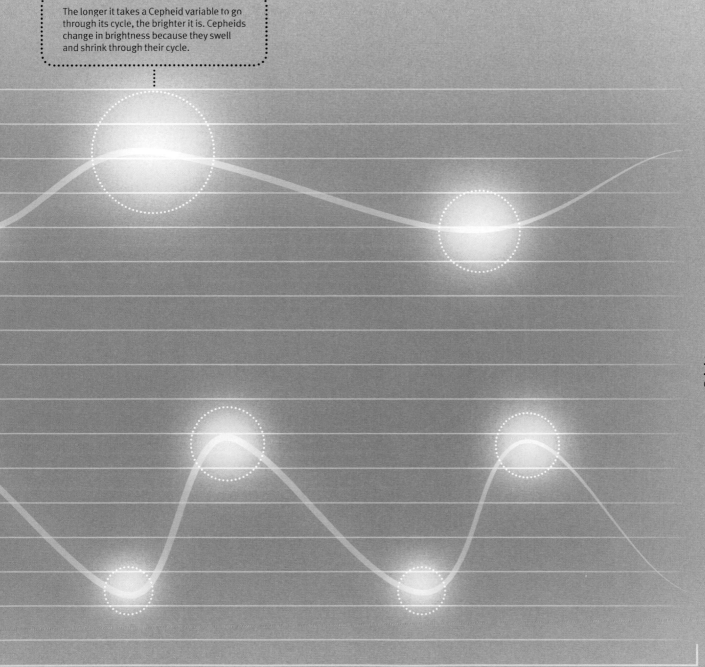

Island Universes_

The Great Debate was the title of a controversy in the world of astronomy in the 1920s. The American astronomers Harlow Shapley and Heber Curtis were on opposite sides of the debate. Did the Milky Way represent the whole of the Universe, as Shapley argued? Or were some of the fuzzy patches of light called 'nebulae' actually 'island universes' – systems like the Milky Way but far beyond it – as Curtis claimed?

The earliest theory of the nebulae after their discovery was that they were the birthplaces of stars. This is true of some of them, but not of others. For example, in 1917 Heber Curtis had noticed 11 novae, or exploding stars, in photographs of the Great Nebula in Andromeda. But these all looked very faint. If in reality they were no fainter than other novae, then the Great Nebula must be very far away.

Armed with Henrietta Leavitt's new way of measuring the distances of stars with Cepheid variables (see pages 60–61), Edwin Hubble, at the observatory on Mount Wilson, California, measured the distances of stars in dozens of nebulae. He found them to be at truly enormous distances, millions of light years away. Others had made similar claims before, but Hubble had done the most solid work to date.

Clearly the Great Nebula in Andromeda would have to be renamed the Great Galaxy in Andromeda, since it must be a system of stars, gas and dust like our own Milky Way. Once the right relationship between the period of a Cepheid and its brightness had been sorted out some decades later, the Andromeda Galaxy was found to be over two million light years from the Milky Way.

Something to Think About . . .

When he first succeeded in measuring the distance to the spiral in Andromeda, Hubble sent a letter to Harlow Shapley. Shapley remarked to a colleague who was in his office when he first read it: 'Here is the letter that has destroyed my universe.'

Andromeda Galaxy
Edwin Hubble showed that the large 'nebula' in the constellation of Andromeda was not a relatively nearby gas cloud but a vast and distant collection of stars, gas and dust like the Milky Way.

The Expanding Universe_

Edwin Hubble progressed from calculating how far away galaxies are to studying how they move, as revealed in their spectra. When the light of an astronomical object is converted into a spectrum, there are usually gaps in the spectrum band – some colours are missing from the light, with the gaps appearing as dark lines or bands crossing the spectrum. Sets of lines are recognizable as the chemical signatures of particular atoms or molecules.

Hubble studied the dark lines in the spectra of galaxies and found that nearly all are shifted towards the red (longwave) end of the spectrum. The wavelengths of the light had been stretched before reaching us, and therefore the positions of the gaps in the spectrum had shifted too. Light waves would be stretched like this if the galaxy were rushing away from us. Hubble claimed that this was the explanation, though other scientists had other ideas. The astronomical world agreed, after some fierce debate, that Hubble was right.

Hubble found that the further away a galaxy is, the faster it is receding. If a galaxy is at twice the distance of a second one, it is retreating at twice the speed of that second one. The faintest and furthest galaxies he could measure were rushing away at a large fraction of the speed of light.

Something to Think About . . .

If you could view the Universe from another galaxy, you would see our Milky Way travelling away from you. And all the other galaxies would appear to be receding with greater speeds if they were further away. At any given time the Universe looks the same from any viewpoint. The expansion has no centre.

Hubble's Law

The further away a galaxy is, the faster it is travelling away from us.

Top: Galaxies 9 billion light years away recede at 66 per cent of the speed of light.

Middle: Galaxies 6 billion light years away recede at 44 per cent of the speed of light.

Bottom: Galaxies 3 billion light years away recede at 22 per cent of the speed of light.

Note: When a galaxy is described here as, say, 3 billion light years away, this is a shorthand way of saying that the light by which we see it has been travelling 3 billion years to reach us. The galaxy itself is now further away.

02.13 Radio Astronomy: Tuning in to the Sky_

Our atmosphere allows radio waves to pass through it freely, just like light waves (see pages 58–59). When radio communication was developed, it became possible to look through this radio 'window' to observe an unseen universe. In 1933 an American radio engineer working for Bell Telephone Labs, Karl Jansky, discovered that a steady radio hiss was coming from the direction of the centre of the Milky Way, our own galaxy. In 1937 an Illinois engineer, Grote Reber, built an instrument in his backyard dedicated to receiving radio waves from space – the first radio telescope. It was a tilting bowl-shaped reflector that focused radio waves from a specific direction to a receiver mounted at the focus.

After the Second World War, newly developed radio and radar technology was applied to this new field of radio astronomy. Some radio telescopes were structures of wire mesh; some were large arrays of rod-shaped aerials; and some were steerable dishes like Reber's visionary design. In 1957 the newly built steerable radio dish at Jodrell Bank in Cheshire, 76m (250ft) in diameter, became famous when it tracked the first artificial satellite, the Soviet Sputnik 1.

Radio astronomers went on to discover pulsars, map the spiral arms of the Galaxy, probe the Galaxy's core and observe the outpourings of energy from quasars and active galaxies. Signals from radio telescopes in locations across the globe, and even in orbit, can be combined to create images as detailed as would be provided by a single instrument as big as the Earth – or even bigger.

Square Kilometre Array

Construction is due to begin on the Square Kilometre Array in 2016. When complete it will consist of thousands of telescopes scattered over thousands of kilometres. Due to be located in either Australia or South Africa, the telescopes will have a combined surface area of 1 sq. km ($\frac{1}{3}$ sq. mile). The flood of data will need the equivalent of a hundred million laptops to process.

Very Large Array

The VLA (Very Large Array) is a huge system of 27 linked steerable radio telescopes mounted on railway tracks in the New Mexico desert. The 27 telescopes are arranged in the shape of the letter Y. The arms of the Y are 21km (13 miles) long.

The Arecibo telescope

The world's largest radio dish is built into a smoothed-out volcanic crater at Arecibo in Puerto Rico. The dish is 305m (1,000ft) wide and is fixed: it scans a strip of the sky as it is carried round by the daily rotation of the Earth.

Something to Think About . . .

We don't just have to receive radio from the sky – we can also send radio signals. Radar astronomy tracks meteors by bouncing radar pulses from the glowing trails of meteors – even in daylight. And radar pulses have been bounced from the rings of Saturn – a round trip of over two hours. The radar echoes give information about the composition and size of the particles making up the rings.

Monster Telescopes_

The biggest telescope that Edwin Hubble ever used was the 5.1m (200in) Hale Telescope (the 200 inches referred to the diameter of the telescope's mirror). In the six decades since then, astronomy in visible light (which now goes hand-in-hand with infrared astronomy) has become vastly more powerful.

⋯⋗ Mirrors are even larger: the mirror of the Japanese Subaru telescope on Mauna Kea in Hawaii is 8.2m (27ft) across.

⋯⋗ Adaptive optics is used to alter the shape of large telescope mirrors, perhaps hundreds of times per second, to cancel out the twinkling of sky objects caused by air turbulence.

⋯⋗ Telescopes are perched on high mountains, above most of the haze and pollution of the atmosphere.

Telescopes in space

Telescopes are now linked to sharpen the images they produce: for example, the Keck Telescope on Mauna Kea in Hawaii combines the light of two 10m (33ft) mirrors 85m (279ft) apart.

Many telescopes have gone into space. The Hubble Space Telescope's successor, the James Webb Space Telescope, will be placed far beyond the Moon, its super-cooled 6.5m (21ft) mirror protected by a giant sunshade.

Mauna Kea
Hawaii

Something to Think About . . .

Human observers never sit at the eyepieces of the largest astronomical telescopes. Cameras are connected to the eyepieces, building up images over many hours, sometimes over many nights. The astronomer may control the telescope from the other side of the globe.

The European Extremely Large Telescope (E-ELT)

The planned E-ELT will be a super-instrument located on a Chilean mountain peak. Its mirror will be 42m (138ft) across comprising 1,000 segments.

Mirrors can be made even larger by building them in segments: for example, the GTC (Gran Telescopio Canarias) in the Canary Islands has a 36-segment mirror 10.4m (34ft) across. Several times per second the individual segments are realigned by individual motors – a technique called adaptive optics.

Gran Telescopio Canarias

Canary Islands

Neutrino Detectors: Astronomy From Underground_

Observatories have not just climbed the highest mountains in recent decades – they have also gone underground into deep mines. Shielded by miles of rock from extraneous radiation, mines are perfect places to observe not light but neutrinos. These ghostly particles are emitted in unimaginable numbers every second from the nuclear reactions at the heart of the Sun and every other star. The neutrino is incredibly unsociable – it can slip through light years of dense matter with only a 50 per cent chance of reacting – but a minute fraction of them can be observed reacting here on Earth.

Among the quadrillions upon quadrillions that pass through the liquid filling the huge underground chamber of a neutrino observatory, a tiny number will react, producing a flash of light that can be detected by photo-detectors.

In 1987 neutrino detectors around the world detected a burst of neutrinos – 24 in all. Three hours later, their source was seen in visible light – a supernova (a gigantic stellar explosion) in the Larger Magellanic Cloud, a nearby galaxy.

Something to Think About . . .

Billions of neutrinos are streaming through your body at this moment. But there's no need to worry: you are almost completely transparent to them, so they almost never react with the atoms in your body. In your lifetime, only a couple of neutrinos will have their paths bent by interactions with you.

This subterranean neutrino telescope is a ball 12m (39ft) across, floating in a water-filled chamber 2km (1¼ miles) underground in a Canadian mine. When a neutrino interacts with a molecule in the 780 tonnes of liquid inside the sphere, a tiny flash of light is registered by one of the 9,600 detectors around the outside of the ball.

Neutrinos are constantly produced in the heart of the Sun, along with photons ('particles' of light and other kinds of electromagnetic radiation). The photons take a million years to struggle to the surface as they constantly interact with matter on the way. Neutrinos speed out of the Sun in about two seconds.

For over 30 years the neutrinos detected from the Sun posed a mystery: there were less than half as many as expected. We now know that the Sun produces the expected number of neutrinos, but some of them alter as they travel, so that our detectors do not register them.

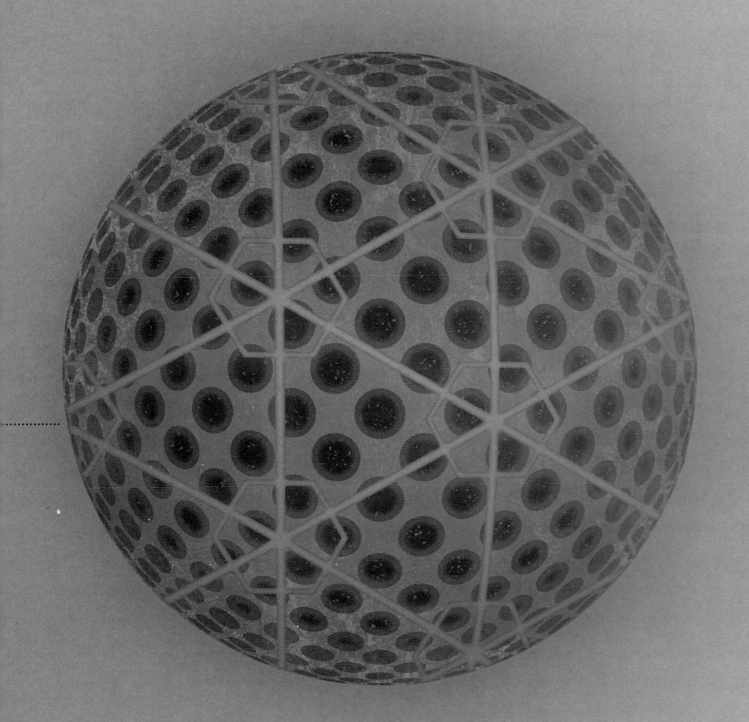

Astronomical Time Machines_

Plans are afoot to build the most gigantic astronomical 'telescope' ever – consisting of orbiting space detectors positioned at the corners of a space triangle measuring 5,000,000km (about 3,000,000 miles) along each side. And all to detect signals from space that have never been unambiguously seen, but which scientists are convinced must be there.

Gravity waves are shivers in space and time predicted by Einstein's General Theory of Relativity. These should be produced by any system in which extremely large masses are being violently accelerated – dying stars collapsing at the ends of their lives, companion stars spiralling inwards towards mutual annihilation, black holes swallowing stars, and so on.

The idea is that the spacecraft will orbit the Sun, sharing the Earth's orbit but 50,000,000km (about 3,000,000 miles) behind, with laser beams linking them. When gravity waves pass, the ripple will affect the length of the beams by less than the diameter of an atom – but that will be detectable.

Something to Think About . . .

It's believed that some gravity waves may have come from the Big Bang itself – so this gravity-wave telescope would be a time machine giving us a view closer to creation than any telescope yet built.

A future space-based gravity-wave detector would consist of three spacecraft millions of kilometres apart, connected by laser beams. Gravity waves spreading out from violent events across the Universe would cause tiny distortions in the beams.

Although gravitational waves have not been detected, scientists are certain they exist. One reason is that the radio signals from some pulsars (extremely accurate astronomical 'clocks' – see pages 168–69) are slowing down, presumably as energy is carried away by gravitational waves.

Some physicists have suggested that we might be able to detect gravitational waves not just from the Big Bang – but even from *before* it (see pages 214–15).

Chapter 03.0

Probes, Satellites & Spaceships_

Into the Unknown_

Understanding the Universe better meant getting beyond the Earth's
dirty, swirling atmosphere to get a clearer view of the stars, and travelling
further on to get a close-up look at the Moon and planets. Only rockets
could provide propulsion beyond the atmosphere. From the 1960s human
beings have escaped the bonds of Earth in hops that gradually grew larger,
passing milestones that had been set earlier by planes and balloons.

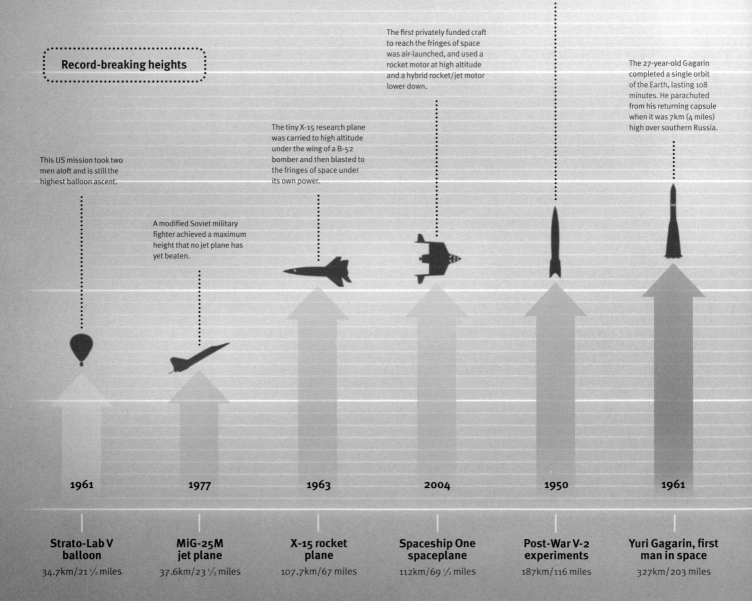

Record-breaking heights

The USA used some of the
Nazis' rocket bombs, which they
had captured before the end
of the Second World War, for
scientific and military research.

The first privately funded craft
to reach the fringes of space
was air-launched, and used a
rocket motor at high altitude
and a hybrid rocket/jet motor
lower down.

The 27-year-old Gagarin
completed a single orbit
of the Earth, lasting 108
minutes. He parachuted
from his returning capsule
when it was 7km (4 miles)
high over southern Russia.

The tiny X-15 research plane
was carried to high altitude
under the wing of a B-52
bomber and then blasted to
the fringes of space under
its own power.

This US mission took two
men aloft and is still the
highest balloon ascent.

A modified Soviet military
fighter achieved a maximum
height that no jet plane has
yet beaten.

1961	1977	1963	2004	1950	1961
Strato-Lab V balloon	**MiG-25M jet plane**	**X-15 rocket plane**	**Spaceship One spaceplane**	**Post-War V-2 experiments**	**Yuri Gagarin, first man in space**
34.7km/21 ½ miles	37.6km/23 ⅓ miles	107.7km/67 miles	112km/69 ½ miles	187km/116 miles	327km/203 miles

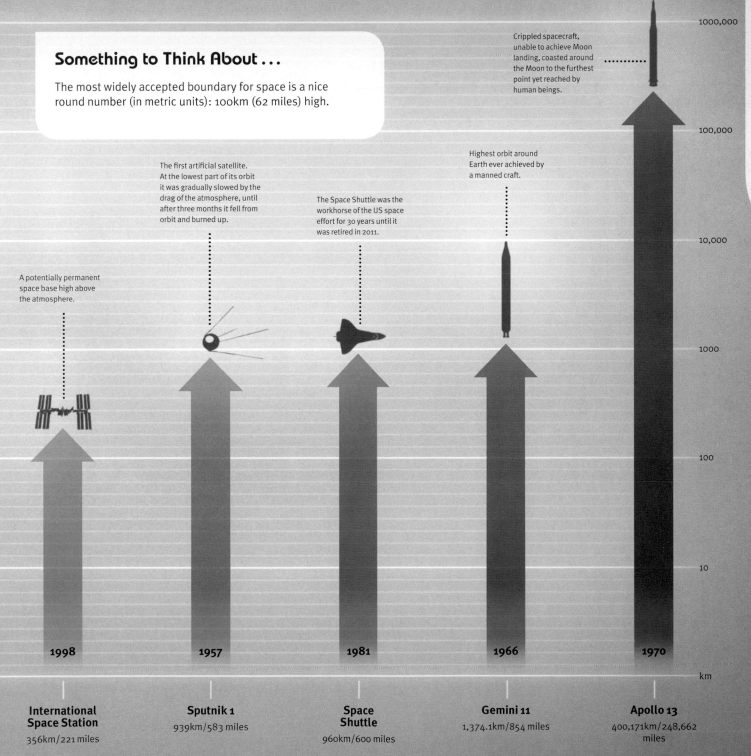

Something to Think About . . .

The most widely accepted boundary for space is a nice round number (in metric units): 100km (62 miles) high.

Crippled spacecraft, unable to achieve Moon landing, coasted around the Moon to the furthest point yet reached by human beings.

The first artificial satellite. At the lowest part of its orbit it was gradually slowed by the drag of the atmosphere, until after three months it fell from orbit and burned up.

The Space Shuttle was the workhorse of the US space effort for 30 years until it was retired in 2011.

Highest orbit around Earth ever achieved by a manned craft.

A potentially permanent space base high above the atmosphere.

1,000,000

100,000

10,000

1000

100

10

km

1998

1957

1981

1966

1970

International Space Station

356km/221 miles

Sputnik 1

939km/583 miles

Space Shuttle

960km/600 miles

Gemini 11

1,374.1km/854 miles

Apollo 13

400,171km/248,662 miles

03.2 Space Stations_

Once astronauts had made brief forays beyond the atmosphere and quickly returned to the safety of the ground, it was time to establish more permanent settlements in space. A true space station would provide a home for a succession of astronauts.

The Soviets set the pace in long-duration space flights. Salyut 1 was a craft that stayed in orbit for a mere six months in 1971, and was occupied for just three weeks of that period. A series of craft followed, until Salyut 7 was launched in 1982 and stayed in orbit for over eight years, hosting Soviet cosmonauts and some foreign guests. It was followed by Mir, which was in orbit from 1986 to 2001; new parts were added until it consisted of seven modules. Several cosmonauts remained on Mir for over a year.

In 1998 cosmonauts began to assemble the International Space Station whilst in orbit. It has been continuously occupied since 1 October 2000 and completion of the station is scheduled for 2012. However, the prospects for keeping the station going for many years beyond that are uncertain.

Something to Think About . . .

The US space station Skylab, consisting of a Saturn rocket stage converted to a laboratory, came to the end of its working life in 1979 after six years of service. It wasn't disassembled before re-entering the atmosphere, with the result that chunks of it hit the Earth in south-western Australia.

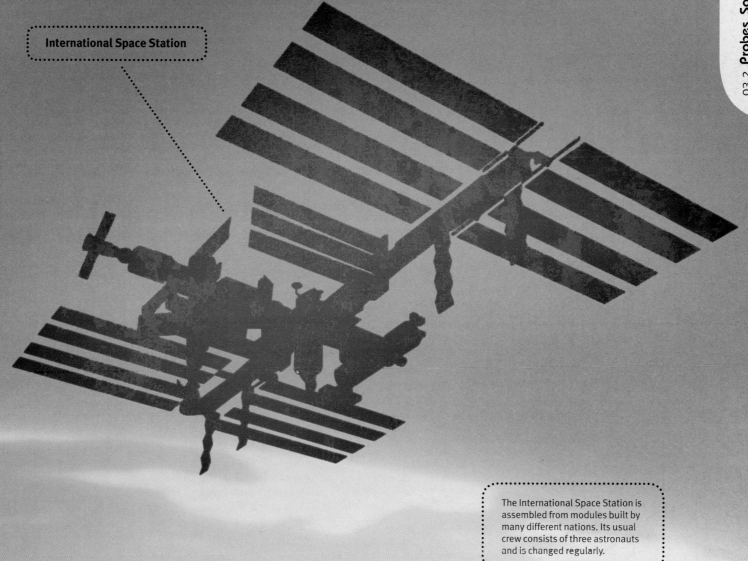

International Space Station

The International Space Station is assembled from modules built by many different nations. Its usual crew consists of three astronauts and is changed regularly.

03.3 # One Giant Leap_

The USA had lagged badly in space achievement as the Soviets claimed a wide range of firsts, but in 1961 President John F. Kennedy declared that by the end of the decade America would land a man on the Moon, and return him to Earth safely.

On 16 July 1969 a 111m (364ft), 3,000-tonne Saturn V rocket thundered into the Florida sky on the Apollo 11 mission. The tiny module that touched down on the Moon's Sea of Tranquillity four days later weighed 14.7 tonnes. The Command Module that splashed down in the Pacific Ocean four days after that weighed just 6 tonnes.

Six Apollo missions followed, five of them successful.

Something to Think About . . .

The Apollo missions brought back a total of 382kg (842lb) of rocks from the Moon. The Soviets also brought samples back, using robot spacecraft – 320g (11oz) or less than one-thousandth as much.

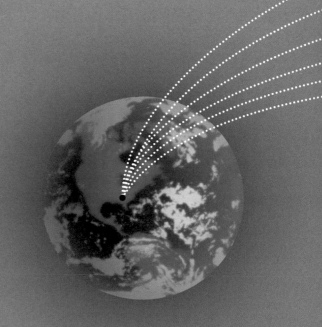

Apollo 12

14 November 1969
Brought back pieces of
the Surveyor unmanned
probe that had made a hard
landing two years earlier.

Apollo 11

16 July 1969
First manned landing
on the Moon.

Apollo 13

11 April 1970
Near-disaster when an onboard
explosion crippled the power
supplies. The landing was called
off but the ship was miraculously
brought safely back to Earth.

Apollo 14

31 January 1971
Commander Alan Shepard
played two golf strokes
on the Moon.

Apollo 17

7 December 1972
Longest stay on Moon's
surface – just 20 seconds
short of 75 hours.

Apollo 16

16 April 1972
First mission to land in
the lunar highlands.

Apollo 15

26 July 1971
Lunar Roving Vehicle, or
'Moon buggy', used.

Destination Mars_

Nobody foresaw, after the last Apollo Moon landing in 1972, that more
than four decades would go by without any human beings returning to
the Moon. Plans for the next great goal – a mission to Mars – have blown
hot and cold. At present there is a tentative plan by the USA, declared in
2007, to send a mission to Mars. The European Space Agency also has
plans to send human beings to Mars some time after 2030. Other
superpowers may have similar plans but these have not been made public.

Impossibly large amounts of fuel would be needed for a return trip from
the Earth to Mars. So some mission ideas involve setting up a manned
Moon base and building rockets there that could make the trip to Mars.

Another plan involves sending an automated chemical factory to Mars.
When it had manufactured enough fuel, astronauts would follow, and they
could use the fuel for their return.

No other planets in the Solar System are realistic for human habitation.
The giant outer planets are balls of gas. A base on Mercury's barren surface
would be scorched half the time, frozen the rest. On the surface of Venus
the hot, dense atmosphere would roast and crush any inhabitants. But
there are many satellites on which colonies could be established, for
those tempted by scenes of rock and ice ...

Day 1

Day 500

Day 1
Take off from Earth
Enter Earth orbit
Leave Earth orbit

Day 223
Enter Mars orbit

Day 234
Landing on Mars

Day 245
Return to Mars orbit

Day 252
Leave Mars orbit

Day 500
Land on Earth

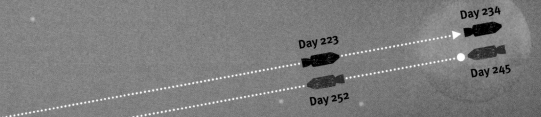

2030 European Space Agency
proposed mission to Mars

Day 234

Day 223

Day 245

Day 252

Something to Think About...

The Apollo Moon 'bases' were disappointingly temporary.
One way to ensure a permanent colony on Mars is described
by various 'Mars to Stay' initiatives: the basic idea being
to send 15 couples to Mars, in five expeditions – on a
one-way trip.

03.5 Robot Explorers_

While the human race has been moving slowly beyond the confines of the Earth, its robot explorers have been forging ahead.

Hundreds of satellites crowd around the Earth, watching our weather, carrying our television programmes, phone conversations and Internet traffic, and mapping Earth's resources. Probes have flown by all the planets and most of the satellites in the Solar System, and in some cases have landed. The frontrunners are the Voyager probes launched in 1977, which will probably cross the Solar System's official interstellar boundary in 2015.

Spacecraft need energy to do this, and it comes either from solar or nuclear energy. Solar energy is harvested by large 'sails' that are unfolded as soon as the craft is outside the atmosphere and free of its launch vehicle. But beyond Mars sunlight is feeble, and so small nuclear generators are often used instead. A mass of radioactive material, usually plutonium-238, constantly generates heat, which is turned into a few hundred watts of electricity.

The craft's lifetime is likely to be limited by the fuel supplies for the 'attitude' (orientation) controls. When the fuel runs out, the craft will no longer be able to keep its radio aerial directed towards the Earth, and contact will be lost.

Something to Think About . . .

When a radio command is sent to a probe making complex manoeuvres near Saturn, it takes over an hour to get there. A controller on Earth would have to wait two to three hours to find out whether the command had been successfully carried out, and what the next command should be. That's why space probes have to be smart robots, capable of deciding on actions and carrying them out without waiting for instructions from Earth.

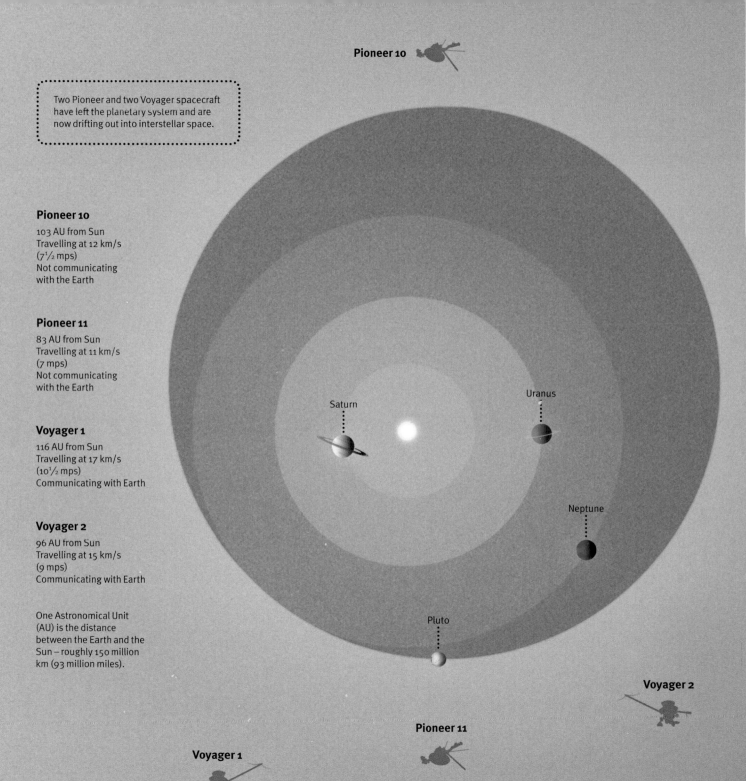

Pioneer 10

Two Pioneer and two Voyager spacecraft have left the planetary system and are now drifting out into interstellar space.

Pioneer 10

103 AU from Sun
Travelling at 12 km/s
(7½ mps)
Not communicating
with the Earth

Pioneer 11

83 AU from Sun
Travelling at 11 km/s
(7 mps)
Not communicating
with the Earth

Voyager 1

116 AU from Sun
Travelling at 17 km/s
(10½ mps)
Communicating with Earth

Voyager 2

96 AU from Sun
Travelling at 15 km/s
(9 mps)
Communicating with Earth

One Astronomical Unit
(AU) is the distance
between the Earth and the
Sun – roughly 150 million
km (93 million miles).

Saturn

Uranus

Neptune

Pluto

Voyager 2

Pioneer 11

Voyager 1

85

03.6 Slingshot Space Travel_

Space probes would need enormous amounts of fuel to travel to the outer reaches of the Solar System if they relied only on their own rocket engines to battle against the Sun's powerful gravitational pull. Voyages to the outer planets are made feasible by the 'slingshot', or gravity-assist, technique.

The craft swings by a planet, which is itself moving around the Sun, and the planet whirls the craft round rather as a thrower whirls a stone around in a sling (think David and Goliath!). The craft is flung on into space with a boost to its speed of up to twice the planet's orbital speed, depending on the precise geometry of the encounter. The first missions beyond the asteroid belt – the two Voyager missions, launched in 1977 – used a rare alignment of the outer planets. Voyager 1 was boosted by flybys of Jupiter and Saturn; Voyager 2 did the same with Jupiter, Saturn, Uranus and Neptune.

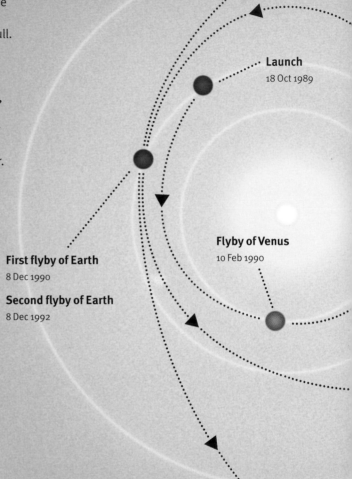

Launch
18 Oct 1989

Flyby of Venus
10 Feb 1990

First flyby of Earth
8 Dec 1990

Second flyby of Earth
8 Dec 1992

Something to Think About . . .

One day in the future long-lived space colonies could wander around the Solar System following convoluted paths called the Interplanetary Transport Network. They would need very little fuel because gravity would do all the work, but it would take many centuries to drift from the neighbourhood of one planet to another.

Encounter with asteroid Gaspra
29 Oct 1991

Path of the Galileo Mission
The Galileo space probe was launched in October 1989 and after more than six years of travel, including three gravity assists from Venus and Earth, reached the giant planet Jupiter.

Arrival at Jupiter
7 Dec 1995

Mars

Encounter with asteroid Ida
28 Aug 1993

Galileo Path
- Earth
- Venus
- Mars
- Asteroids
- Jupiter

Living in Space_

The time must come when the first true space colonies – fully self-supporting settlements beyond the atmosphere – are established. Once they exist, the human race would be able to survive whatever catastrophes occur back on Earth. Space colonies might be space stations circling the Earth, another planet or a satellite, or they might be cities on other planets or satellites.

In 1974 an American physicist, Gerard K. O'Neill, outlined his idea for enormous space colonies using technologies little more advanced than today's in his paper entitled 'The Colonization of Space'. One design featured two parallel cylinders, 32km (20 miles) long and 6.5km (4 miles) wide. Inside there would be land and water, with a daily cycle of light and darkness controlled by shades over windows running the length of the cylinders. Several million people would live on board.

We're a long way from creating such structures, but we've passed the first milestones. It will be another significant time when the first human being is born in space, and another when the first human beings spend their whole lives in space.

Something to Think About . . .

'In the long run, a single-planet species will not survive. One day, I don't know when, but one day, there will be more humans living off the Earth than on it.'
(Former NASA director Mike Griffin)

Biosphere colonies

We can imagine setting up colonies on any satellite in the Solar System, and any planet with a solid surface (the rocky inner planets). But they all have environments hostile to life. Colonists would have to live inside biospheres, where temperature and atmosphere would be controlled and where the food to sustain life could be grown.

Reaching for the Stars_

Flying to other stars will be vastly harder than colonizing the Solar System. If the distance from the Sun to the Earth were scaled down to 1m (3¼ft), the distance to the furthest planet, Neptune, would be about 30m (98ft) – but the distance to the nearest star, Proxima Centauri, would be 270km (168 miles).

The Voyager 1 space probe is travelling out of the Solar System at about 17 km (10½ miles) per second, one and a half times as fast as an Apollo Moon-rocket; but this is a crawling pace that means it would take a whopping 76,000 years to get as far as Proxima Centauri.

According to modern physics, the limiting speed for any object is the speed of light, 300,000km (186,000 miles) per second. A few ways to get a starship to approach this speed would be:

· ·> H-bomb power: A series of hydrogen-bomb explosions behind a ship would blast it through space.

· ·> Antimatter: Can only be made in tiny quantities in particle accelerators like the Large Hadron Collider. When antimatter meets ordinary matter they annihilate one another producing energy. If we could make antimatter in large quantities, it would make a powerful fuel for starships.

· ·> Light sail: Rather than fuel, the ship would carry a huge 'sail' that is pushed by a powerful light beam from an Earth-based laser.

Something to Think About . . .

Interstellar space is not completely empty. A starship travelling at tens of thousands of kilometres per second would have to be armoured against the dust grains, gas and 'cosmic rays' (subatomic particles) that would batter it.

Enormous amounts of energy would be needed to drive a starship at near-light speeds, yet journeys to most stars would take centuries or millennia.

Future fuel

A theoretical method of powering spacecraft
was proposed in 1960 by the American physicist
Robert W. Bussard. The Bussard ramjet is a type
of fusion rocket that would use electromagnetic
fields to compress the hydrogen of the incredibly
thin interstellar gas, creating fuel that could be
'burned' in a thermonuclear reaction. Bussard's
idea has proved immensely popular in science-
fiction novels, films and TV shows.

03.9 Colonizing the Galaxy_

If we want to find other Earth-like planets in the Galaxy, to act as new homes for the human race, we need to send starships across the interstellar wastes. There are at least 200 billion stars in the Galaxy, and many, perhaps most, have planets circling them. Exploring this colossal number of worlds might seem an impossible task, but in the 1970s the Scottish writer Chris Boyce popularized a technique for exploring the Galaxy in a reasonable time. It depends on building robot space probes that are self-replicating – able to build exact copies of themselves. They are called von Neumann probes after the Hungarian-American mathematician John Von Neumann, who studied the concept of self-replicating machines.

··‑⟩ Probes – say, just two of them – are launched towards nearby star systems. The most promising target stars might be spaced an average of 100 light years apart, depending on how choosy we are. At a cruising speed of one tenth of the speed of light, it would take typically 1,000 years for each ship to get to its target.

··‑⟩ A probe is launched towards a nearby star system. If the target is 100 light years away and the probe cruises at one tenth of the speed of light, the craft will take 1,000 years to get there.

··‑⟩ On arriving, the probe mines the planets and asteroids of the star system to build two replicas of itself. These are then launched towards further target stars to repeat the process. The most promising stars might be spaced an average of 100 light years apart, depending on how choosy we are in selecting targets.

··‑⟩ If there are suitable planets, a probe can plant human colonies, using frozen embryos or completely synthetic cells.

··‑⟩ After 20,000 years and 20 generations the furthest probes will now be 2,000 light years from Earth – and there are about a million of them.

··‑⟩ The furthest reaches of the Galaxy are about 100,000 light years from the Earth. The last probes will get there in a million years.

Something to Think About . . .

Many people are appalled by this plan, regarding it as akin to infecting the Galaxy with a plague. If the project were ever launched it would be necessary at least to program the probes to wind down the exploration after a set number of years had passed.

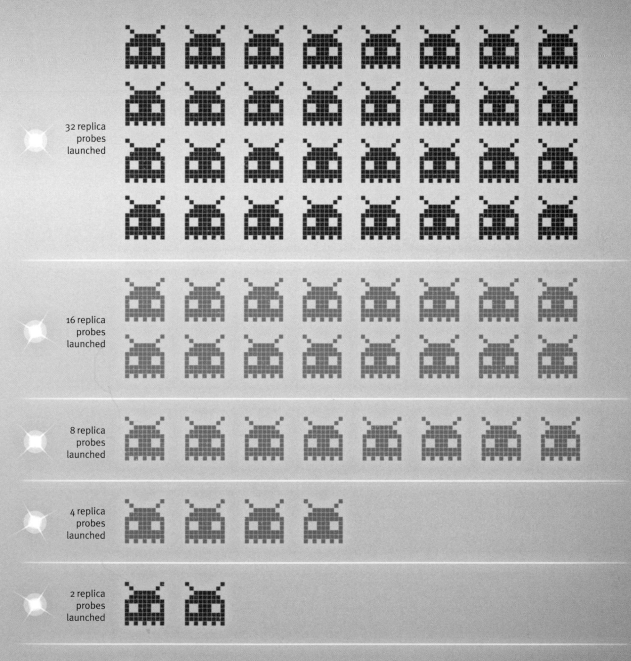

32 replica probes launched

16 replica probes launched

8 replica probes launched

4 replica probes launched

2 replica probes launched

  Probe launch from Earth

Self-replicating probes could spread from star to star across the Galaxy.

Chapter 04.0

The Sun & the Rocky Planets_

Inhabitants of the Solar System_

The Solar System is divided into several major districts, each with its own characteristic type of inhabitant. The easiest unit to use for marking out these regions is the Astronomical Unit (AU), the distance between the Earth and the Sun – about 150 million km (93 million miles).

⋯⇢ The Sun lies at the centre, keeping everything else in its gravitational grip and providing heat and light for the whole system.

⋯⇢ Four small planets, made of rock, the Earth among them, lie within 2AU of the Sun.

⋯⇢ The asteroid belt is a wide belt of small rocky objects, the asteroids, lying 2–3.5AU from the Sun.

⋯⇢ The four gas giants consist of huge globes of gases including hydrogen, helium, methane and ammonia, hiding hot liquid or solid Earth-sized cores. They lie between 5 and 30AU from the Sun.

⋯⇢ A belt of small objects stretches beyond Neptune, the furthest planet from the Sun. The largest of these is Pluto, long regarded as a planet.

⋯⇢ The Oort cloud, an enormous shell of small objects made of ice and rock, lies unobserved, as much as halfway to the nearest stars. We know of it only because comets occasionally fall from it into the inner Solar System.

⋯⇢ Some comets, asteroids and other small bodies wander into the inner Solar System and out again.

Something to Think About . . .

The Sun accounts for 99.9 per cent of all the mass in the Solar System, and 90 per cent of the rest belongs to Jupiter and Saturn.

All distances shown are measured from
the Sun. This illustration is not to scale.

Kuiper belt

30–55AU

Asteroid belt

Earth

1AU

Pluto

39AU

**Rocky
planets**

**Gas
giants**

Sun

Oort cloud

5,000–100,000AU

Astronomical Units

One Astronomical Unit (AU) equals 149,597,871km
(92,955,807 miles). The distance between the
Earth and the Sun is approximately 1AU.

97

The Solar System: a Family Portrait_

The members of the Solar System exhibit a vast range of sizes. The Sun, its senior member, is 1.39 million km (865,000 miles) across – nearly 300 times the diameter of diminutive Mercury, the smallest planet.

To make some more comparisons: Jupiter is roughly 11 times the diameter of Earth, so 1,400 Earths could fit inside Jupiter. The Sun is ten times the diameter of Jupiter, and over 900 Jupiters, or 1.2 million Earths, could fit into the Sun.

The sizes and compositions of the planets are so hugely different because of the processes of the birth of the Solar System billions of years ago (see pages 100–101). As they exist today, the only planets other than our own Earth on which human beings could comfortably set foot, would be Mars and the cooler polar regions of Mercury.

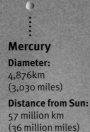
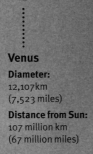
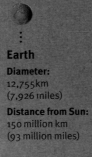
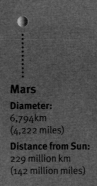

Mercury
Diameter:
4,876km
(3,030 miles)

Distance from Sun:
57 million km
(36 million miles)

Venus
Diameter:
12,107km
(7,523 miles)

Distance from Sun:
107 million km
(67 million miles)

Earth
Diameter:
12,755km
(7,926 miles)

Distance from Sun:
150 million km
(93 million miles)

Mars
Diameter:
6,794km
(4,222 miles)

Distance from Sun:
229 million km
(142 million miles)

Jupiter
Diameter:
142,983km
(88,846 miles)

Distance from Sun:
777 million km
(483 million miles)

Something to Think About . . .

Although the planets shown here are to scale, the distances between them are not. These two pages would need to be over 1km ($^5/_8$ mile) wide to show the distance between the Sun and Neptune, the furthest planet. Even the distance between the Sun and Mercury would be 18m (59ft).

Saturn

Diameter:
120,536km
(74,898 miles)

Distance from Sun:
1,429 million km
(888 million miles)

Uranus

Diameter:
51,117km
(31,763 miles)

Distance from Sun:
2,871 million km
(1,784 million miles)

Neptune

Diameter:
49,527km
(30,775 miles)

Distance from Sun:
4,496 million km
(2,794 million miles)

Sun

Diameter:
1.39 million km
(865,000 miles)

Birth of the Solar System_

The Solar System we see today was born 4.5 billion years ago from a cold, dark cloud of hydrogen mixed with some other gases, and with dust, floating between the stars in a Galaxy that was already eight billion years old.

- The cloud, several light years across, collapsed under its own gravitational force, spinning faster as it did so and becoming a rotating disc. The core of the cloud was heated by energy released by the collapse, becoming a hot and glowing protostar.

- In the swirling cloud, dust began to gather into denser clumps, which grew into chunks of rock and metal.

- After 50 million years, thermonuclear reactions in the Sun 'switched on', and it became a true star.

- Solid chunks in the cloud continually collided and a few built up into planets. Light materials such as gaseous hydrogen, helium, water and methane were driven out of the hotter inner parts of the cloud.

- In the cooler outer regions, the chunks of rock, ice and other frozen materials grew large and gathered vast volumes of gas around themselves.

- Beyond the orbit of Neptune, matter was too thinly dispersed to gather into planets.

- Small rocky bodies that might have formed a fifth inner planet were prevented from doing so by Jupiter's strong gravity. They remained as the asteroid belt.

- Because of complex gravitational interactions, the gas giants swapped places repeatedly, flinging debris from the inner Solar System beyond the orbits of the planets.

Something to Think About . . .

The collapse of the primordial gas cloud was probably triggered when a nearby star exploded as a supernova, compressing the gas and kick-starting the collapse.

The primordial Sun glows at the heart of a swirling cloud of gas and dust. Heat and pressure have triggered thermonuclear reactions in the Sun.

Inner Solar System: lighter gases will be driven from this warmer region, and the planets that form here will be mostly rock.

Outer Solar System: in this cooler region small rocky bodies will build up to form embryo planets, which will gather gases to become the gas giants.

101

The Face of the Sun_

For the last 4.5 billion years the Solar System has been presided over by a Sun that seems calm and stable. But the telescope reveals that the apparently unchanging Sun has a turbulent face.

The surface of the Sun is mottled with granules, about 1,000km (621 miles) across, where gas is rising and falling. Dark spots appear, grow in numbers and size, and then decrease again in an 11-year cycle. When the bright disc of the Sun is hidden behind the Moon in a solar eclipse (see page 114), a bright red rim of glowing hydrogen called the chromosphere ('colour sphere') can be seen. Prominences, bright red filaments or loops, can also be observed leaping out from the Sun.

Corona
The Sun's rarefied, faint, but intensely hot outer atmosphere.

Something to Think About . . .

If the total energy output of the Sun for one second could be collected, it could power the entire USA, at its present rate of consumption, for the next nine million years.

Flare
An explosion of hot gas that sends waves of charged particles out through the Solar System.

Warning!
Never look at the Sun through binoculars or a telescope. You can use these instruments to project an image of the Sun on to paper or card that you can view safely. It is also not safe to look at the Sun through sunglasses or dark film – only use a filter designed to block infrared and ultraviolet radiation.

Prominence

A glowing column of gas rising from the surface, formed from an expanding loop of gas where magnetic fields 'leak' out of the Sun.

Chromosphere

A layer of glowing red hydrogen gas.

Sunspot

Areas that are relatively cool and strongly magnetized, often larger than the Earth.

Granules

Regions of the Sun where hot, low-density solar material comes to the surface. The material rises from the depths at the centre of each granule, cools, becomes denser, and sinks outside the granule.

Anatomy of the Sun_

The Sun changed from a protostar to a true star when thermonuclear reactions began in its core (see pages 100–101). The Sun then consisted of about 74 per cent hydrogen and 24 per cent helium with a smattering of other elements. In the hot, dense core of the gas cloud, the atoms of these elements were broken down, so that they consisted of the nuclei, or cores, of the atoms swimming in a sea of electrons.

The hydrogen nuclei collided and welded together to form a new kind of nucleus – helium – while throwing out energy as electromagnetic radiation. The Sun had begun a ten-billion-year career of generating energy by turning hydrogen into helium – it's now about halfway through ...

Something to Think About ...

In the core of the Sun 620 million tonnes of hydrogen are converted into helium every second. However, 4.3 million tonnes disappear completely, having been changed into energy.

Diameter
* 1,392,000km (865,000 miles)
* 109 x Earth

Distance from Earth
* 149,597,871km (92,955,807 miles)

Rotation period
* 25.05 days at equator
* 34.4 days near poles

Mass
* 1.9891 x 10^{30} kg
* 1.9891 billion billion billion tonnes
* 330,000 x Earth

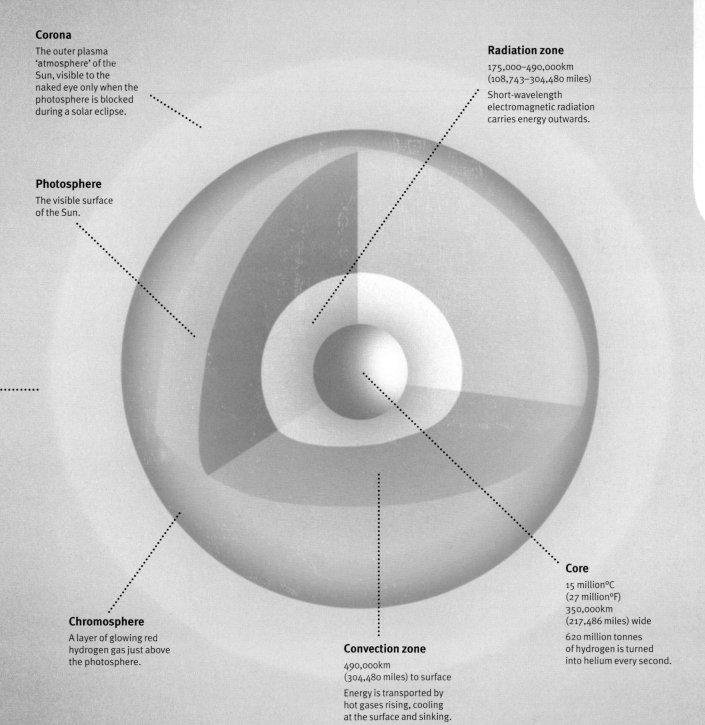

Corona

The outer plasma 'atmosphere' of the Sun, visible to the naked eye only when the photosphere is blocked during a solar eclipse.

Radiation zone

175,000–490,000km (108,743–304,480 miles)

Short-wavelength electromagnetic radiation carries energy outwards.

Photosphere

The visible surface of the Sun.

Core

15 million°C (27 million°F) 350,000km (217,486 miles) wide

620 million tonnes of hydrogen is turned into helium every second.

Chromosphere

A layer of glowing red hydrogen gas just above the photosphere.

Convection zone

490,000km (304,480 miles) to surface

Energy is transported by hot gases rising, cooling at the surface and sinking.

Mercury: the Scorched Planet_

The innermost planet of the Solar System looks a lot like the Moon. Its airless, arid surface is peppered with craters formed four billion years ago, when billions of small rocky bodies still circulated in the Solar System and bombarded the newly formed planets. Once Mercury may have been much larger, but its outer layers were stripped off by a collision with an unusually large body. That could explain why it has a large core extending to three-quarters of the radius of the planet, consisting largely of iron. There are no very large plains like those on the Moon.

Because Mercury is close to the Sun, we see it only briefly after sunset or before dawn. It is best placed for observation from Earth every 116 days – nearly equal to two rotation periods. This means that it always presents the same face to us at these times. This resulted in astronomers mistakenly believing its rotation period was 88 days – equal to its 'year' – rather than 58.6 days – two-thirds of the 'year'. They discovered the truth when they bounced radar pulses off the planet in 1965.

Something to Think About . . .

Even on Mercury there could be supplies of water for future visitors from Earth. Ice survives in the craters near the poles that are permanently shaded from sunlight.

Mass
* 330.22 billion billion tonnes
* 0.055 x Earth

Diameter
* 4,879km (3,032 miles)

Rotation period ('day')
* 58.6 days

Distance from Sun
* 57.9 million km (36.0 million miles)

Orbital period ('year')
* 88 days

Number of moons
* 0

Mercury's orbit puzzled generations of astronomers: it is an ellipse, in agreement with Kepler's First Law (see pages 46–47), but an ellipse that slowly rotates around the Sun. Explanations included an undiscovered planet even closer to the Sun and disturbing Mercury's orbit. There's no such planet – Einstein explained the effect in 1916 with his General Theory of Relativity.

The Sun

Venus: the Planet from Hell_

Venus at its brightest is a brilliant jewel suspended in the morning or evening sky, so bright it can cast a shadow. The clouds that totally cover the planet are actually responsible for its brightness. In the past even serious astronomers thought those clouds could hide hot, wet jungles. Venus might once have had oceans of water, but it is too hot now for water to exist.

The surface hidden by the clouds is a roasting 470°C (880°F), hot enough to melt lead. And those dazzling white clouds consist of sulphuric acid droplets – if there's any rain on Venus, it consists of sulphuric acid, and it evaporates before it reaches the ground. The atmosphere consists almost entirely of carbon dioxide and there is so much of it that the pressure at ground level is a crushing 92 times that on Earth. The resulting greenhouse effect makes Venus hotter than Mercury.

Something to Think About . . .

When astronomers first bounced radar pulses off the surface of Venus, they were astonished to discover that the planet rotates in the opposite (retrograde) direction compared to the other planets, and takes about 243 of our days to do so.

Years of radar mapping by the orbiting Magellan probe revealed the surface of Venus that is perpetually hidden by clouds. Although this is a false-colour image, the chosen colour is quite close to the reddish hue of Venusian rocks. The lighter tones are higher land. The long, light-coloured area across the lower part of the image is Aphrodite Terra, an extensive highland region running along the equator.

Mass
* 4,870 billion billion tonnes
* 0.815 x Earth

Diameter
* 12,104km (7,521 miles)
* 0.95 x Earth

Rotation period ('day')
* 243 days

Distance from Sun
* 108.2 million km (67.2 million miles)

Orbital period ('year')
* 224.7 days

Number of moons
* 0

04.8 Earth: No Planet Like Home_

The Earth exists in a 'Goldilocks' zone of the Solar System that is neither too close to the Sun, nor too far from it, so that the surface, like Baby Bear's porridge, is neither too hot nor too cold for liquid water to exist.

The planet also has internal heat. Some of this is left over from the heating effect caused by the proto-Earth condensing from the primordial gas cloud, but most of it is due to the small proportion of radioactive minerals scattered through the planet. This internal heat means that there is a hot liquid core (actually the outer core – the innermost core is solid), which acts like a dynamo creating the Earth's magnetic field. The heat also keeps the overlying rocky layer, called the mantle, churning like soup heated in a pan. This churning drives the constant reshaping of the Earth's crust, on which we live.

Something to Think About . . .

The most common element in the Earth is iron (32 per cent), closely followed by oxygen (30 per cent). These are combined with other elements in the rocky mantle and crust, but the core is almost pure iron, with some nickel. In the atmosphere, the most common element is nitrogen, comprising 78 per cent of the atmosphere by numbers of molecules. Oxygen only accounts for 21 per cent of the atmosphere.

Mass

✷ 5,974 billion billion tonnes

Diameter

✷ 12,756km (7,926 miles) equator
✷ 12,714km (7,900 miles) polar

Rotation period (day)

✷ 23.93 hours

Distance from Sun

✷ 149.60 million km
✷ 92.96 million miles

Orbital period (year)

✷ 365.26 days

Number of moons

✷ 1

Atmosphere

No definite upper boundary: 90 per cent of its mass lies below 16km (10 miles).

Crust

Depth:
0–6okm
(0–37 miles)

Plates of light rock float on the fluid hot rock beneath. The crust is thin where it forms the floor of ocean basins, thicker where it forms the continents.

Inner core

Depth:
5,150–6,36okm
3,200–3,952 miles

With a diameter of 2,400km (1,491 miles), the inner core consists of iron that is hotter than the surface of the Sun, crushed into solid form by the pressure of the overlying mass.

Oceans

Depth:
10.9km
(6¾ miles)

This surface ocean of liquid water is unique in the Solar System.

Mantle

Depth:
35–2,890km
(22–1,796 miles)

A rock layer in which convection currents of hot material are constantly rising, cooling and sinking, driving continental drift.

Outer core

Depth:
2,890–5,150km
(1,796–3,200 miles)

Molten iron and nickel. Currents in the outer core generate the planet's magnetic field.

111

Earth: the Living Planet_

Many things combine to make Earth a home that is conducive to life:

⋯→ The atmosphere keeps the surface at a warm temperature that makes it possible for water to stay liquid, providing a transport medium for the tissues of living things.

⋯→ The Earth's magnetic field shields the surface from high-energy particles from the Sun and the Galaxy beyond, and prevents the atmosphere from being stripped away.

⋯→ The dynamic surface, shaped by the convection movements in the mantle, absorbs carbon dioxide from the atmosphere, keeping temperature and pressure moderate.

⋯→ The ozone layer, high in the atmosphere, shields organisms from most of the damaging ultraviolet radiation in sunlight.

The oxygen in the atmosphere is highly reactive. It's the Earth's plants that first created the oxygen-rich atmosphere and now constantly replenish it. If astronomers detect oxygen-rich atmospheres on planets around other stars, it will be a strong signal that there is life on those worlds.

Something to Think About . . .

The Earth was not born with all the water that it now possesses. A large part of it was brought to the young Earth by impacts of icy bodies such as comets, from the outer regions of the Solar System.

Within the first billion years of the Earth's existence, the first complex self-reproducing molecules formed whether in the deep sea, a lake, or perhaps in icy tundra, is not known. Reproduction with variation, together with natural selection by the environment, led to a proliferation of evermore complex forms.

04.10 The Moon: Our Night Light_

The 'inconstancy' of the Moon is due to the changing relationship between the Moon and the direction of the Sun. We see the changing phases of the Moon as we see more or less of the lit-up portion of the Moon. For example, we see the Moon as full when it is opposite the Sun and the side facing us is illuminated.

After the full moon it takes approximately 27 days 8 hours for the Moon to move once around the Earth in relation to the stars. But during that time the Moon and Earth travel approximately $\frac{1}{12}$th of the way around the Sun, and it takes a bit longer for the Moon to return to its position opposite the Sun. So the Moon goes through all its phases in approximately 29 days 13 hours.

Sometimes at full moon all or part of the Moon enters the Earth's shadow. This is a lunar eclipse.

Something to Think About . . .

By coincidence the Moon occasionally looks almost exactly the same size in the sky as the Sun. Sometimes at new moon the Moon moves exactly between the Earth and the Sun. Where the Moon's shadow falls on the Earth, the Sun is wholly or partly obscured for a short period in a solar eclipse.

Mass
* 73.5 billion billion tonnes
* 0.0123 x Earth

Diameter
* 3,476km (2,160 miles)

Rotation period = orbital period
* 27.3 days

Distance from Earth
* 384,400km (238,855 miles)

Number of moons
* 0

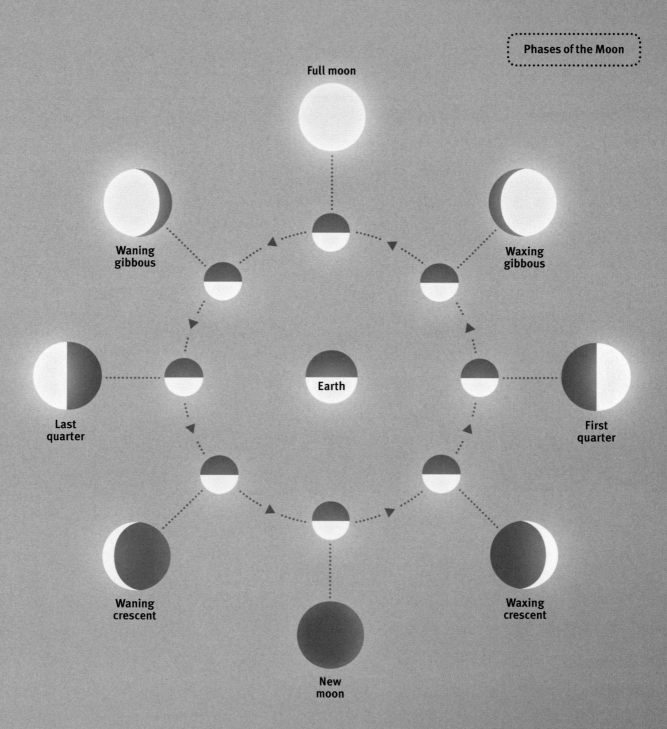

04.11 The Moon: the Earth's Companion_

The Moon is an arid, airless, lifeless wasteland, scorched by the Sun during the two-week lunar day, and chilled to minus 170°C (minus 275°F) during the equally long lunar night. The dark patches we can see with the naked eye are called *maria* (singular *mare*), the Latin for 'seas', and that is in fact what they once were – seas of molten lava, which have now solidified into plains of rock.

When the *maria* erupted from the Moon's interior, they covered what was there before. Elsewhere, in the lunar highlands, that older landscape survives. It is covered by thousands upon thousands of craters, making it resemble an enormous battlefield. The craters were created by the great bombardment of small rocky bodies that took place when the Solar System was in its infancy. Light-coloured rays of ejected material, hundreds of kilometres in length, stretch away from some craters.

On the far side of the Moon, unseen by human beings until Apollo 8's first circumnavigation in 1968, there are very few *maria*. The whole landscape is heavily cratered.

Something to Think About . . .

Eugene Cernan was the last man to walk on the Moon, on 19 December 1972, at the end of the Apollo 17 mission. But there is still someone up there – another Eugene. Some of the ashes of the American astronomer Eugene Shoemaker (1928–1997) were crashed on to the Moon by the Lunar Prospector probe in July 1999.

Birth of the Moon

The Moon was born 4,500 million years ago – according to the current, most-favoured theory. A body about the size of today's Mars collided with the young Earth. Wreckage from the body and from the Earth's outer layers was flung into space. Some never returned and some fell back to Earth, but some material gathered in a ring orbiting the Earth, and then collected to form the Moon. This explains why the Moon resembles the Earth's outer rocks.

Mars: the Red Planet_

Every two years Mars comes its closest to Earth and becomes a bright, baleful, red 'star' glowing in the night sky. During the 19th century the telescope revealed it as a world that was very Earth-like in some ways: it had bright white polar caps that shrank with the Martian summer, while dark markings elsewhere grew, suggesting they might be vegetation.

The notion caught on that Mars was a 'dying' planet, once as lush as the Earth, but now a desert. It has been the scene of countless imaginary adventures, such as the *Mars* stories by the creator of *Tarzan*, Edgar Rice Burroughs. And it has been the base from which countless attacks against Earth have been launched, as in H. G. Wells's classic *The War of the Worlds*, published in 1898. The fiction writers had to adapt when space probes discovered the true nature of the planet (see pages 120–121), but Mars is still a favourite scene for their imaginings.

A suitable case for treatment

Scientists have suggested that Mars could be terraformed – made comfortable for human beings. Various ways of providing the planet with a dense, warm atmosphere of oxygen plus a good supply of water have been suggested. There is a lot of frozen carbon dioxide at the poles, which could warm the planet if released.

Something to Think About . . .

The Italian astronomer Giovanni Schiaparelli published maps of Mars in which he included fine, straight lines that he believed were 'channels' – the Italian word for which is *canali*. He thought water flowed from the icecaps in summer, causing vegetation to grow in areas nearer the equator. Plenty of other observers thought they could see the lines. Some – notably the American astronomer Percival Lowell – decided the channels were actually canals, constructed by intelligent beings.

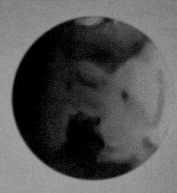

Mars today

Mars is cold, dry and hostile to life. Its red deserts consist of reddish iron-rich rocks – essentially, rust.

Bringing rain

The Martian atmosphere is warmed – perhaps using rocket-loads of CFC greenhouse gases, or perhaps by scattering soot at the poles to absorb sunlight, warm the ground and release frozen-in carbon dioxide. The warming would force water vapour into the atmosphere to fall as rain.

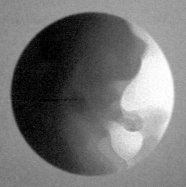

The greening of the planet

As the ocean basins fill, plants from Earth can thrive on the warmer, moister planet.

A new Earth

After thousands of years, human beings and animals cloned from Earth stocks can live on the surface of the transformed planet.

04.13 Mars in Reality_

When the first close-up photos of Mars were taken by the US Mariner 5 probe in 1965, astronomers were not surprised to see that there were no canals. But they were astonished to find the southern hemisphere covered with craters. Invisible as they were in the fuzzy images of Mars seen in Earthbound telescopes, almost no one had predicted them.

··❯ The atmosphere of Mars is composed almost entirely of carbon dioxide. It is exceptionally thin – in the lowest-lying areas of the surface, where the pressure is highest, it is only about one per cent of the pressure of Earth's.

··❯ The ice caps are water ice. The southern ice cap is covered with a permanent thick layer of frozen carbon dioxide. The northern ice cap has a thinner coating of carbon dioxide in winter, which sublimes (evaporates without becoming liquid) during summer.

··❯ More than a quarter of the atmosphere solidifies on the winter pole and then sublimes again, so the atmospheric pressure across the planet varies wildly. The thin air nevertheless whips up enormous dust storms that can blot out detail over the globe.

··❯ An enormous system of canyons called the Valles Marineris ('Valleys of Mariner', named after the space probe that discovered them) run a quarter of the way around the planet's equator.

··❯ Mars boasts the Solar System's tallest known mountain, Olympus Mons. It rises 22,000m (72,160ft) above the surrounding landscape, whereas Mount Everest, the highest mountain on Earth, rises 8,848m (29,021ft) above sea level. But Olympus Mons is so broad and its slopes are so shallow that anyone at its foot or on its summit would get little sense of its height.

The surface of Mars
The surface of Mars consists of reddish, iron-rich rocks and sand. There are many craters but they have been eroded over the millennia by the action of winds. The surface is etched in many places with gullies apparently made by flowing water.

Mass
* 642 billion billion tonnes
* 0.107 x Earth

Diameter
* 6,792km (4,220 miles)

Rotation period ('day')
* 24.62 hours

Distance from Sun (average)
* 228 million km (142 million miles)

Orbital period ('year')
* 687 days

Number of moons
* 2

Something to Think About . . .

Mars has two tiny satellites, Deimos ('Terror') and Phobos ('Fear'). The inner one, Phobos, whizzes round Mars in 7 hours 39 minutes, much faster than the planet rotates, so from the surface of Mars Phobos would seem to rise in the west and set in the east. Jonathan Swift eerily predicted the existence of two Martian moons in *Gulliver's Travels* in 1726. In Gulliver's third voyage he visits Laputa, a flying island ruled by mad scientists, who give him (not very accurate) details of Martian moons that were not to be discovered in reality until 1877.

Chapter 05.0

Planetary Giants & Dwarfs_

Jupiter: Monarch of the Planets_

The huge mass of the largest planet, Jupiter, dominates the Solar System. Jupiter's gravity prevented a planet from forming in the large gap between Jupiter and Mars (see pages 56–57). It influences the motion of the next planet out, Saturn (which has 30 per cent of Jupiter's mass): Jupiter revolves around the Sun five times while Saturn revolves twice.

Three-quarters of Jupiter's mass is hydrogen; most of the rest is helium, with a little ammonia and methane and traces of many other substances. The planet's rapid spin gives it a noticeable bulge at the equator, and it drives the cloud systems into bands of light and dark cloud parallel to the equator.

Inside Jupiter, temperatures and pressures rise with depth. The bulk of the interior is hydrogen crushed into a metal-like state, and at the centre of the planet there is probably a solid core, perhaps of hydrogen mixed with rock.

Something to Think About . . .

In 1994 Comet Shoemaker–Levy 9 crashed into Jupiter. Although the impact was on the far side of Jupiter, where it could not be seen from Earth, it was observed by the Galileo space probe. As Jupiter rotated, markings could be seen in the atmosphere where the comet had hit, which lasted for months.

Mass
* 1.899 million billion billion tonnes
* 318 x Earth

Diameter (equator)
* 142,984 km (88,846 miles)

Rotation period ('day')
* 9.92 hours

Distance from Sun
* 778.5 million km (483.7 million miles)

Orbital period ('year')
* 4,331.6 days

Number of moons (at least)
* 64

The enormous globe of Jupiter, ten times broader than the Earth, is banded with bright zones and dark belts of cloud.

The Great Red Spot
The Great Red Spot is a storm larger than the Earth that has been raging in Jupiter's southern hemisphere for centuries.

Satellite and shadow
A moon passes between Jupiter and the Sun, its shadow passing across Jupiter's face.

Jupiter: a Miniature Solar System_

Jupiter has at least 64 satellites – the list constantly lengthens as tiny new ones are identified. Jupiter has undoubtedly had many more satellites in the past, which it has swallowed while greedily dragging new small passers-by into its extended family.

The innermost eight satellites were born alongside Jupiter in the Solar System's primordial cloud of gas and dust. Four of these are tiny, rocky inner satellites, analogous to the Solar System's rocky inner planets. The next four satellites out from the planet are Jupiter's largest. The closest of these, Io, is constantly being stretched and squeezed as it passes through different regions of Jupiter's gravity, causing it to heat up and volcanoes to form. Europa and Ganymede, the next large satellites, are heated less strongly by gravitational stresses, but the heating is enough to cause their cores to melt. Callisto, the outermost of the large satellites, experiences little gravitational heating.

Jupiter's other satellites are small, distant bodies that formed elsewhere in the Solar System and were captured by the planet. They move in orbits that are tilted, elongated and sometimes retrograde – moving the 'wrong' way around Jupiter.

Something to Think About . . .

Jupiter has faint, dark rings, not discovered until 1979 when the Voyager 1 space probe got a close-up view. They are formed by dust being blasted from the surface of the small innermost moons by meteoric impacts.

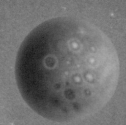

Io

Io is the smallest of the satellites discovered by Galileo, the closest to Jupiter, and possibly the most active body in the Solar System – apart from the Sun. The sulphur spewed on to its surface from volcanoes colours it reddish yellow.

Europa

The smooth surface is water ice, criss-crossed with cracks. Warmer currents rising from below churn the surface and erase any impact scars.

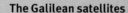

The Galilean satellites

The four largest and brightest of Jupiter's satellites were discovered in 1609 and 1610 when at least two scientists turned telescopes on to the sky for the first time. One was Simon Marius, a German astronomer. The other was Galileo Galilei, the Italian genius who used his results in his battle for a new world view and a new physics, and for whom the satellites have been named.

Ganymede

The largest moon in our Solar System. Ganymede is bigger than the planet Mercury. It consists of ice and rock, but it has a magnetic field, probably generated in a molten core.

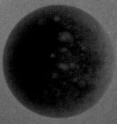

Callisto

Callisto is a dark ball of ice and rock, about the same size as Mercury. Its surface has not been reworked by volcanic activity, and is scarred by ancient impact craters.

Saturn: the Ringed Planet_

The most beautiful sight through the astronomer's telescope must be Saturn. Its broad, bright rings consist of orbiting ice particles, ranging in size from grains to chunks the size of a small car. Saturn itself spins so fast that it has a notably oblate or 'squashed' look, its polar diameter being nearly ten per cent less than its equatorial diameter.

The average density of Saturn is actually less than that of water, but there is a dense, rocky core at the planet's hot centre. The gaseous body of Saturn is about 90 per cent hydrogen by mass, and most of the rest is helium. Storms are visible as light-coloured elliptical spots called ovals, and once in every Saturnian 'year' (29.5 Earth years), a large storm in the form of a Great White Spot develops – a rather poor rival to Jupiter's Great Red Spot (see page 125).

Something to Think About . . .

When Galileo first observed Saturn through his telescope in 1610, he could see there was something strange about the planet – he thought that it was a triple body, or else a single object with 'ears'. In 1612, however, the rings were edge-on to Earth and the 'ears' were no longer visible. The next year they reappeared. Galileo was baffled. Only in 1755 did the Dutch astronomer Christiaan Huygens realize Saturn is surrounded by rings.

Mass
- ❋ 568.5 thousand billion billion tonnes
- ❋ 95.1 x Earth

Diameter (equator)
- ❋ 120,536km (74,900 miles)

Rotation period ('day')
- ❋ 10.7 hours

Distance from Sun
- ❋ 1,433.5 million km (890.8 million miles)

Orbital period ('year')
- ❋ 10,759 days

Number of moons (at least)
- ❋ 62

05.4 Saturn: Ringlets & Braids_

Space scientists were stunned by the first close-up pictures of Saturn's rings taken by the Voyager probes in the early 1980s. No one had foreseen their incredibly intricate structure. Some questioned whether there was some unknown flaw in our understanding of gravitation.

⋯⋗ The broad rings seen from Earth are made of thousands of thin rings, dubbed ringlets.

⋯⋗ There are transient kinks in some of the ringlets, giving them a braided look.

⋯⋗ Spokes appear and disappear in the ring.

⋯⋗ Faint, thin rings were discovered; that wasn't too surprising, but sometimes they were incomplete, forming arcs.

We've come to understand them better now, and we haven't had to throw out Newton's theory of gravitation.

Saturn's satellites are crucial in shaping the rings:

⋯⋗ Sometimes moons force gaps in the rings. For example, the Cassini Division is controlled by the satellite Mimas, which lies beyond the rings. If a particle strayed into the gap, it would orbit twice every time Mimas orbited once, and would be disturbed so that it was driven out of that orbit.

⋯⋗ Satellites within the rings sweep out gaps: for example, the moon Pan orbits in the Encke Gap.

⋯⋗ Electrical forces seem to be responsible for holding bunches of particles above the rings to form spokes.

⋯⋗ Ripples and waves are also created in the rings by satellites as they pass.

The rings may be the debris of a satellite that was smashed by an impact or torn apart by Saturn's gravity when it wandered too close to the giant planet. No one is sure whether this happened when the Solar System was young, or much more recently.

Encke Gap

The moon Pan orbits within this gap, sweeping it clear of particles. Nevertheless, there are several ringlets within the gap.

G Ring

This faint, thin ring contains a section, or arc, that is denser than the rest of the ring, centred on a tiny moonlet called Aegaeon.

Cassini Division

The gravitational influence of the moon Mimas, orbiting much further out, keeps this region clear of particles.

C Ring

This tenuous, dusky band has been called the Crepe Ring from its appearance.

A Ring

Outermost of the three rings discovered from Earth. The outer edge is sharply defined by the effects of two small satellites just outside the ring, Janus and Epimetheus.

D Ring

The D Ring lies inside the C Ring and is even dimmer. It consists largely of dust, and extends downwards to the top of the atmosphere.

B Ring

The broadest and brightest of the rings also contains the most mass. Seen close up, it looks 'grooved' like a vinyl LP.

F Ring

The thin F Ring is held in place by two 'shepherd' moons, Prometheus and Pandora, which move on each side of the ring.

E Ring

The vast and diffuse outermost ring (much further out than the edge of this page) consists of fine grains from 'ice volcanoes' on the moon Enceladus.

Something to Think About . . .

If all the material in Saturn's rings were gathered into one mass, it would probably make a ball smaller than the tiny moon Mimas, which is about 400km (249 miles) across.

131

Saturn's Moons_

Over 60 satellites swarm around Saturn – not including the countless icy grains and chunks that make up its rings, and an unknown number of moonlets, or mini-moons, up to about 400m (¼ mile) in diameter, which move within the rings and create much of their complex structure.

Saturn's system of rings and moons has been tirelessly explored by the Cassini probe since its arrival in 2004, after over six years of travel from Earth. The highlight of the mission was sending the Huygens lander to parachute through the thick atmosphere of the largest moon, Titan. The lander sent back pictures of a rugged mountainous landscape as it descended. After landing it showed the ground around the landing-site, strewn with 'pebbles', probably made of ice.

Something to Think About . . .

When Giovanni Cassini discovered Iapetus, the third largest of Saturn's satellites in 1671, he was puzzled that he could only see it on one side of Saturn – when it moved round to the other side, it vanished. He realized that the side of Iapetus that trails in its orbital motion is bright, while the leading side is covered with darker material, either from inside the moon or swept up from space.

Enceladus

Saturn's sixth largest moon is an icy but active world whose surface is carved with grooves and cliffs. 'Ice volcanoes' frequently erupt, suggesting there may be liquid water, perhaps even containing life, beneath the surface.

Dione

Dione is a ball of ice with a rocky core. Geological activity has created towering ice cliffs in places, hundreds of metres high, some of them apparently freshly made.

Titan

The second largest satellite in the Solar System (Jupiter's Ganymede is the largest), Titan's diameter is 40 per cent of Earth's. It has a dense atmosphere of nitrogen with some methane, made opaque to visible light by orange smog. There are also lakes of liquid methane.

Phoebe

The smashed landscape of Phoebe reveals its history of impacts. The satellite revolves around Saturn in a retrograde direction because it was 'captured' after forming elsewhere in the Solar System.

Uranus: the Tilted Planet_

Uranus, the first planet to be discovered in historical times (see pages 54–55), lies as far beyond Saturn as Saturn is distant from the Sun. It is a greenish-blue gas giant that seems to roll along on its side. Astronomers believe a collision with an Earth-sized object pushed the planet onto its side when the Solar System was young. The result is that each pole points to the Sun for half of the Uranus 'year' of 84 years. During this time that half of the planet enjoys its summer – albeit a summer in which the Sun is 1/400th as bright as it appears on Earth.

Uranus is smaller than Jupiter or Saturn, a mere four times the diameter of the Earth. It owes its colour to a haze of methane in the upper atmosphere. The deep gaseous atmosphere is similar to Jupiter's and Saturn's in that it comprises mostly hydrogen and helium, but contains more water, ammonia, methane and other compounds. There is a rocky, hot core.

Uranus's thin, dark rings were only discovered when they were seen to make stars twinkle as they passed in front of them.

Something to Think About . . .

The rotation of Uranus is retrograde – the opposite way to the planet's revolution around the Sun. When a planet revolves the 'wrong' way, which pole is north? You could say it's the pole at which the planet revolves anticlockwise (as it is for Earth). But the International Astronomical Union has decreed that it's the other pole, because it points towards what we on Earth call the northern half of the sky – the half centred roughly on Polaris, the Pole Star (see pages 26–27).

The rings of Uranus are parallel to the planet's equator, and tilted nearly at right angles to the planet's orbit. They are thin and dim compared with Saturn's rings. Brilliant white clouds of methane crystals dot the bluish-green methane haze on the planet.

Mass
* 86,800 billion billion tonnes
* 14.5 x Earth

Diameter (equator)
* 51,118 km (31,764 miles)

Rotation period ('day')
* 17.2 hours

Distance from Sun
* 2,872.5 million km (1,785 million miles)

Orbital period ('year')
* 84.3 years

Number of moons (at least)
* 27

Neptune: the Farthest Planet_

Neptune appears as a livelier twin of Uranus. It is slightly more massive but, because that extra mass squeezes the interior, the planet is slightly smaller. Like Uranus, Neptune consists mostly of hydrogen and helium in its outer regions, and of a hot core of rock and other compounds. But for unknown reasons Neptune has a more bluish hue than Uranus. Our most detailed view of Neptune has come from the only probe to visit the planet, Voyager 2, which flew past it in 1989.

Neptune was first discovered on paper, and was only later identified with a telescope. An English astronomer, John Couch Adams, and a French mathematician, Urbain Le Verrier, laboured independently to work out the position of an unknown planet that seemed to be causing disturbances in the movement of Uranus. Unknown to each other, they calculated very similar positions for the unknown body. Using Le Verrier's prediction, the German observer Johann Galle discovered Neptune on 23 September 1846, close to the expected position. Adams's prediction was followed up more lethargically by a British observer, James Challis, who missed Neptune even when he had it in his field of view. Eventually Adams won recognition as co-discoverer of Neptune.

Something to Think About . . .

Neptune has five named rings, and some less conspicuous unnamed ones. In the bright Adams ring, there are five arcs, or denser regions. They are long-lived, having changed little since they were first observed by Voyager 2. They are somehow kept in position by the gravitational pull of one or more moons, and are fed with material that is blasted from the moons by meteoric impacts.

The bluish disc of Neptune looms in the gloom of the outer Solar System, where it receives only 1/900th of the sunlight that Earth receives. Dark spots the size of the Earth sometimes appear in the atmosphere of Neptune. They are storms resembling the Great Red Spot of Jupiter, but shorter-lived.

Mass
* 102,430 billion billion tonnes
* 17.1 x Earth

Diameter
* 49,530 km (30,777 miles)

Rotation period ('day')
* 16 hours 7 minutes

Distance from Sun
* 4,503 million km (2,798 million miles)

Orbital period ('year')
* 164.8 years

Number of moons (at least)
* 13

Moons of the Outer Solar System_

Uranus has 27 known satellites and Neptune has 13. Like most of the satellites of Jupiter and Saturn, the moons of Uranus and Neptune have icy surfaces, with the possibility of liquid water oceans lying beneath the surface. Astro-biologists regard these as promising places to seek life.

A tour of the Solar System would find many features among these satellites worth a diversion:

⋯⟶ The surface of Miranda, the smallest of the major moons of Uranus, is mostly ice. There are cliffs there that are 20km (12 miles) high – twice as high as Mount Everest.

⋯⟶ Titania, the largest of the moons of Uranus, consists roughly half of rock and half of frozen materials such as carbon dioxide, water and methane.

⋯⟶ Triton, the largest satellite of Neptune and the second largest in the Solar System, is big enough for its gravity to have pulled it into a spherical shape. Yet it goes around Neptune in the retrograde direction, the opposite to the spin of the planet. Triton was probably born far beyond Neptune's orbit and captured later.

⋯⟶ The inner satellites of Neptune probably did not form together with the planet from the primordial nebula. When Triton was captured it played havoc with the existing system of moons, most of which were thrown into space or pulverized in collisions. The moons we see today were created when the remnants gathered together – rather as our Moon was formed (see page 117).

Proteus
Neptunes's second largest moon.

Uranus

Something to Think About . . .

Perdita is a small inner moon of Uranus whose name is Latin for 'lost one'. The name is appropriate as there was a 13-year gap between it being photographed by Voyager 2 in 1986 and being spotted in the pictures. It was another four years before the Hubble Space Telescope confirmed its existence.

Triton

Characterized by volcanoes of nitrogen and crustal movements. The texture of Triton's surface has been described as 'like a cantaloupe'.

Miranda

The smashed appearance of the surface of this moon is now thought to be due to internal heating by stress as it moves through stronger and weaker regions of Uranus's gravity.

Titania

The icy surface of Titania, largest of the moons of Uranus, is riven by enormous valleys and cliffs. There may be a liquid ocean deep beneath the surface, above a rocky core.

Neptune

Dwarf Planets_

In 2006 the whole world was fascinated when the astronomical powers that be decided to demote Pluto, long regarded as the most remote planet in the Solar System, to the status of a dwarf planet.

The change was prompted by the rapid discoveries of tiny bodies beyond Neptune, which nobody would call planets, yet which seemed to have fundamentally the same nature as Pluto. The term 'dwarf planet' was defined to accommodate these. A dwarf planet:

···⇥ revolves around the Sun as the planets do (not around a larger body as satellites do);

···⇥ has enough mass for its own gravity to pull it into a nearly spherical shape (ruling out many small, irregular-shaped chunks of rock);

···⇥ but is not large enough to have swept its orbit clear of smaller objects (as the eight planets have).

Pluto fits the last condition because most of its elongated, tilted orbit lies in the Kuiper belt, which consists of trillions of objects orbiting beyond Neptune, the great majority of which are too small to count as dwarf planets. The first asteroid to be discovered, Ceres, in the heart of the main asteroid belt between Mars and Jupiter, also fits the definition (see page 56).

Three other bodies in the Kuiper belt have so far been deemed to be dwarf planets, Eris, Makemake and Haumea, although their masses are very uncertain. There could be several hundred more dwarf planets waiting to be discovered in the Kuiper belt, and probably thousands in orbits tilted at angles that keep them away from the Kuiper belt.

Something to Think About . . .

Pluto was discovered in 1930 by a 24-year-old American astronomer, Clyde Tombaugh. He could never have dreamed that he would one day visit Pluto; but that's what he – or rather, 30g (1oz) of his ashes – will do in July 2015, when his remains fly by Pluto in the New Horizons spacecraft, the first probe to visit Pluto.

Makemake

Discovered just before Easter 2005, the third largest dwarf planet was named after a god of the ancient culture of Easter Island.

Haumea

The fourth largest dwarf planet is named after a Hawaiian goddess. It has a highly elongated shape, probably because of a rapid rotation.

Eris

Eris prompted the move to define dwarf planets when it was found that it was probably larger than Pluto. The name appropriately means 'discord'. It is the largest of the dwarf planets.

Pluto

Has only 1/500th of the mass of the Earth and 1/5th of the diameter. It forms a pair with its largest moon Charon, which is about 1/8th of its size. The two bodies orbit around their centre of gravity, which lies in empty space between them.

Ceres

Although having only 1/7,000th of the mass of the Earth, the first asteroid to be discovered was regarded as a planet when it was discovered in 1801, and was nearly reclassified as a planet in 2006.

141

Small Solar System Bodies_

When dwarf planets were defined for the first time in 2006, a ragbag category of leftover bodies was also established. A Small Solar System Body (SSSB) is officially anything in the Solar System that isn't a planet, a satellite, or a dwarf planet. This includes:

⋯▸ the asteroids orbiting between Mars and Jupiter – except Ceres, which is massive enough to have become spherical;

⋯▸ trojans – similar bodies sharing the orbits of Jupiter and some other planets;

⋯▸ comets (see pages 146–147);

⋯▸ centaurs – small bodies moving beyond the asteroid belt and within the orbit of Neptune that can't make up their minds whether they're comets or asteroids.

Undoubtedly some bodies already known and listed as SSSBs will be reclassified as dwarf planets if astronomers discover that they are larger than previously thought.

Something to Think About . . .

The International Astronomical Union has left the minimum size of an SSSB open, so it's unclear whether perhaps even meteoroids count (see pages 148–149).

Uranus

Neptune

The vast majority of SSSBs lie beyond Neptune in the Kuiper belt. But the long-observed asteroids in the main asteroid belt between Mars and Jupiter are also SSSBs, and so are the trojan asteroids, which share the orbits of Jupiter and other planets.

Earth

Jupiter

Main asteroid belt

Mars

Trojan asteroids

Saturn

Too Close for Comfort_

Could a comet or asteroid ever collide with the Earth and cause havoc? In fact, the real question is not 'Could?', but 'When?' There have been many colossal impacts on Earth through the ages. The most famous is the impact 65 million years ago of a body 10km (6 miles) in diameter, which crashed into what is now the Gulf of Mexico, creating gigantic tsunamis, temporarily changing the climate and probably leading to the extinctions of countless species that occurred around this time – notably the dinosaurs.

A body that size arrives less often than once every ten million years. But one object measuring 5–10m (16–33ft) across hits our atmosphere every year on average, releasing the energy of a Hiroshima bomb, though it usually explodes harmlessly in the upper atmosphere.

A number of organizations around the world, under the umbrella title of Spaceguard, are constantly monitoring the sky for near-Earth objects (NEOs). NASA was charged by the US Congress in 2005 with detecting and cataloguing at least 90 per cent of NEOs larger than 140m (460ft) across by 2020. You can check the latest status of the threat to Earth at NASA's neo.jpl.nasa.gov/risk/ site.

The object that received the highest-ever threat rating on the Torino Impact Hazard Scale was asteroid Apophis, which in December 2004, briefly reached level 4. As with every other NEO, improved observations eventually put the threat back to 0.

Torino Impact Hazard Scale
The scale is named after the Italian city of Turin (*Torino* in Italian), where it was presented at a conference. This is an abbreviated version. Find the full version at neo.jpl.nasa.gov/torino_scale

Something to Think About . . .

If you like to mix video-game thrills with bona fide science, you can go to Purdue University's 'Impact: Earth!' asteroid impact effects calculator at purdue.edu/impactearth, where you can set the size, speed and other details of an imaginary NEO and learn just how much mayhem it would create.

No hazard	0	Likelihood of collision effectively zero, or object very small.
Normal	1	A near pass that poses no unusual level of danger. No cause for public concern.
Meriting attention by astronomers	2	A somewhat close pass. No cause for public concern as actual collision very unlikely.
	3	Close encounter with a 1 per cent or greater chance of collision capable of localized destruction. Attention by public officials merited if the encounter is less than a decade away.
	4	Close encounter with a 1 per cent or greater chance of collision capable of regional devastation. Attention by public officials merited if the encounter is less than a decade away.
Threatening	5	Close encounter posing a serious, but still uncertain, threat of regional devastation. More observations needed. If the encounter is less than a decade away, governmental contingency planning may be warranted.
	6	Close encounter by a large object posing an uncertain threat of a global catastrophe. More observations needed. If the encounter is less than three decades away, governmental contingency planning may be warranted.
	7	A very close encounter by a large object, which poses an unprecedented but still uncertain threat of a global catastrophe. International contingency planning warranted.
Certain collision	8	A certain collision, capable of causing localized destruction or a tsunami.
	9	A certain collision, capable of causing unprecedented regional devastation or a major tsunami.
	10	A certain collision, capable of causing global climatic catastrophe that may threaten the future of civilization.

05.12 Comets: Omens in the Sky_

Comets don't shoot dramatically across the sky: they move in a stately fashion against the background of stars as, over a period of months, they fall towards the Sun from the outer Solar System, swing past it, and recede once again. To the naked eye, a comet grows from a point of light to a small fuzzy disc. Then it grows a tail that points away from the Sun, lagging behind on its approach, and leading on its retreat. These tails can be longer than the distance from the Earth to the Sun.

Before and after its trip around the Sun, a comet is just a cold, dark lump, resembling an asteroid (see pages 56–57), except that it also contains icy material. There are trillions of these dark bodies in the Kuiper belt, just beyond Neptune (see pages 142–143), and in the Oort cloud, an invisible reservoir 1–2 light years from the Sun. Sometimes gravitational disturbances push comets out of these locations and into elongated orbits that take them close to the Sun, and that's when they come alive and occasionally become spectacular.

Something to Think About . . .

A comet can make hundreds of passes of the Sun before its icy material is evaporated and the comet has 'worn out'. The rocky remnants may break up, leaving dust spread around the comet's orbit and providing meteor showers for astronomical observers if the Earth happens to cross the orbit.

Dust tail

A curved tail created by dust that is pushed away from the nucleus of the comet by sunlight and the solar wind (a stream of charged particles from the Sun).

Structure of a comet

When it is far from the Sun, a comet is just a ball of rock, mixed with various frozen substances, such as water, carbon dioxide, methane and ammonia.

Coma

A halo of gas and dust thrown off by the comet as the ice of the nucleus warms and evaporates when it is in the vicinity of the Sun.

Nucleus

A ball of rock mixed with icy material that is stirred to activity in the inner Solar System by the warmth of the Sun.

Plasma tail

Electrically-charged particles dragged away from the nucleus by the solar wind, forming a straight tail.

147

05.13 Shooting Stars_

A typical shooting or falling star appears exactly as its name suggests: a point of light that darts across the sky, disappearing a second or two after it appears. It's this visible phenomenon that is known as a meteor. It is created when a grain of dust, a meteoroid, hurtles into the atmosphere, is heated by friction and burns up. Larger fragments of rock can leave a trail of glowing gases that can linger, sometimes for many minutes. Occasionally a meteoroid is large enough for some of the rock to survive the trip through the atmosphere. The pieces that survive the fall are called meteorites.

Some meteors appear as showers, reappearing on predictable dates during the year. This happens when the Earth, in its motion around the Sun, crosses the orbit of a comet, perhaps long disappeared, which has shed dust along its orbit. The meteoroids all move in the same direction when they reach our atmosphere, so if the meteor trails are traced back, they seem to emanate from one point. This point is called the radiant, and the shower is then named after the constellation in which the radiant lies. The brightest annual showers are the Perseids (whose radiant lies in Perseus) and the Leonids (radiant in Leo).

Something to Think About . . .

Of the 50,000 meteorites that have been discovered on Earth, a few dozen are believed to have come from Mars. These are made of rocks much younger than the great majority of meteorites, and their chemical composition matches that of the rocks and atmosphere of Mars. They may have been flung into space by volcanoes, or by the impact of an asteroid with Mars. Doubtless meteoroids flung from Earth reached the Red Planet before the first space probes.

A spectacular meteor 'storm' filled the skies over North America in November 1833. For hours the skies were filled with the fiery streaks, inspiring awe or terror across the country. As the sky turned, the meteors' point of origin turned with them, always remaining in the constellation of Leo. The event persuaded many astronomers for the first time that meteors were caused by objects arriving from outside the atmosphere, and were not purely atmospheric phenomena. When the Leonids put on another spectacular show in November 1866, the Italian astronomer Giovanni Schiaparelli worked out that the objects causing the shower were moving in the same orbit as a newly discovered comet called Tempel–Tuttle.

Chapter 06.0

Stars by the Billion_

Star Light, Star Bright_

In the 2nd century BC, the Greek astronomer and mathematician Hipparchus assigned stars to six classes according to their brightness. He called the classes magnitudes. Stars of the first magnitude were brighter than the second magnitude and so on, down to the sixth magnitude, where the stars are just visible to the naked eye.

The system is still used today, but with modern technology, magnitudes can now be given to decimal places. A star of magnitude 2.3 for example, is defined as being 2.51 times as bright (as measured by instruments) as one of magnitude 3.3, which is 2.51 times as bright as one of magnitude 4.3, and so on. This means that a star is exactly 100 times as bright as another star that is five magnitudes fainter.

Photographic film and electronic detectors used with telescopes are far more sensitive than the human eye, so the magnitude scale has been extended from 6 to beyond 30. The European Extremely Large Telescope, due to go into operation on a Chilean mountaintop soon after 2020, will detect objects as faint as magnitude 36. An object with a magnitude of 36 is a trillion times fainter than the faintest star visible to the naked eye.

Something to Think About . . .

When the magnitude scale was made more precise with the use of photography in astronomy, the four brightest stars had to be given negative magnitudes: Sirius (-1.5), Canopus (-0.7), Arcturus (-0.04) and Alpha Centauri (-0.01).

Apparent magnitudes

How bright objects appear in the sky depends on their actual brightness and how far away they are.

Magnitude range			Number of stars in range
-1.50	┈┈>	-0.51	2
-0.50	┈┈>	+0.49	6
0.50	┈┈>	1.49	14
1.50	┈┈>	2.49	71
2.50	┈┈>	3.49	190
3.50	┈┈>	4.49	610
4.50	┈┈>	5.49	1,929
5.50	┈┈>	6.49	5,946

Counting the stars

An increase in magnitude of 1 means a decrease in brightness by about 2.5 and an increase in the number of stars in the range by about three.

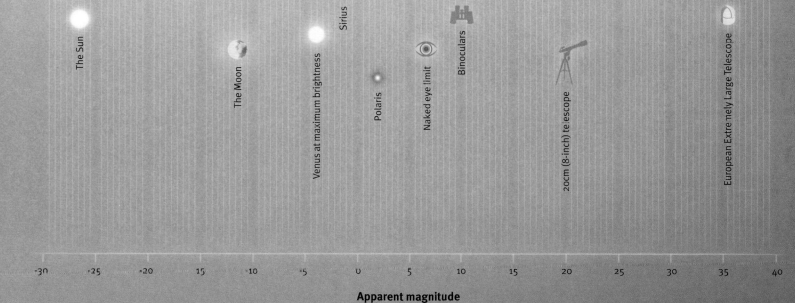

The Sun

The Moon

Venus at maximum brightness

Sirius

Polaris

Naked eye limit

Binoculars

20cm (8-inch) telescope

European Extremely Large Telescope

-30 -25 -20 15 -10 -5 0 5 10 15 20 25 30 35 40

Apparent magnitude

The Stellar Zoo_

Stars are vastly different from one another – in size, temperature, colour and behaviour. These depend mainly on the stage in its life cycle the star has reached – and that in turn depends on how long ago the star was born and how much mass it had when it formed.

A high-mass star burns fast and furiously. Although it has more fuel to burn, it gets through it faster:

⋯⟩ A star ten times as large as the Sun will live a mere 20 million years before it ends its life in a gigantic explosion called a supernova.

⋯⟩ A star with the mass of the Sun will live for ten billion years before running out of hydrogen and swelling to become a red giant. (The Sun is about halfway through that lifetime now.)

⋯⟩ A star born with 6 to 40 per cent the mass of the Sun will last trillions of years (hundreds of times the present age of the Universe), glowing feebly. These objects comprise the great majority of the stars in the Galaxy.

These are the life cycles of solitary stars. As with human beings, a star's life history will become very much more complicated if it spends its life with a partner. Most stars larger than red dwarfs exist in such multiple systems (see pages 164–165).

Something to Think About . . .

The Sun is classified as a yellow dwarf. It's not yellow, it's white, but it emits radiation most strongly in the yellow portion of the spectrum. And in astronomy, any star that's not a giant is a dwarf.

Red supergiant
Betelgeuse

Late stage of a
Sun-sized star

Mass
✳ Up to 10 x the Sun

Size
✳ 300–1,000 x the Sun
(pulsating)

Brightness
✳ 140,000 x the Sun

Temperature
✳ 1,000°C
✳ (1,800°F)

Blue supergiant
Deneb

Late stage of
massive star

Mass
✳ 10–50 x the Sun

Size
✳ 110 x the Sun

Brightness
✳ 50,000 x the Sun

Temperature
✳ 30,000–50,000°C
✳ 54,000–90,000°F

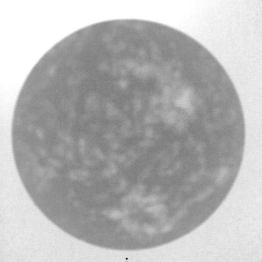

Red giant
Aldebaran

Late stage of a
massive star

Mass
✳ 1.7 x the Sun

Size
✳ 44 x the diameter
of the Sun

Brightness
✳ 425 x the Sun

Temperature
✳ 4,100°C
✳ (7,400°F)

Yellow dwarf
The Sun

Stars with a mass similar
to the Sun. They can be a
range of colours. The Sun
is a typical member of
the class

Mass
✳ 333,333 x Earth

Size
✳ Can hold about a
million planets the size
of Earth

Brightness
✳ Equal to ten trillion
trillion fluorescent light
bulbs

Temperature
✳ Around 6,000°C
✳ (10,800°F)

Red dwarf
Proxima Centauri

The smallest, feeblest
sort of star – and the
most common

Mass
✳ Up to 40 per cent
of the Sun's mass

Size
✳ 30 per cent of
the Sun's diameter

Brightness
✳ Less than $\frac{1}{10}$
of the Sun

Temperature
✳ 3,000°C
✳ (5,400°F)

White dwarf
Sirius B

The collapsed
end-state of a
Sun-sized star

Mass
✳ About the
same as the Sun

Size
✳ The size of
the Earth

Brightness
✳ $\frac{1}{40}$ that of
the Sun

Temperature
✳ 25,000°C
✳ (45,000°F)

06.3 A Star is Born_

Stars are born constantly, and astronomers can watch the process happening. They are formed when a cloud of gas and dust in interstellar space begins to collapse under the influence of its own gravitational attraction. The cloud breaks up into smaller masses, which eventually become individual stars. A cluster of hundreds or thousands of protostars is formed. Our Solar System was formed in just such a cluster (see pages 100–101).

When the Galaxy was young, 13 billion years ago, vast numbers of stars were born from its dense gas clouds. Many of these were in tightly packed globular clusters, each containing thousands of stars (see pages 180–181), distributed through a sphere encompassing the Galaxy. Others were born in smaller open clusters, such as the Pleiades, or Seven Sisters, visible in the constellation of Taurus the Bull. The bright young blue stars of the Pleiades are still surrounded by the wisps of the gas and dust from which they were born.

In today's Galaxy stars are born at the rate of about 100 per century.

Something to Think About . . .

The Orion Nebula is a star nursery that is visible to the naked eye as a fuzzy patch in the constellation of Orion the Hunter, within the 'sword'. In reality, it measures over 20 light years across, and is over 1,300 light years away. It contains thousands of newly forming and young stars, causing the hydrogen gas around them to glow pink in astronomical photographs.

The Pillars of Creation

This star nursery in the Eagle Nebula was photographed by the Hubble Space Telescope. Intense radiation from nearby newly born stars is sweeping the region clear of gas and dust. But at the tip of each 'pillar' is a dark knot of dense gas and dust where new stars are being born. The 'pillars' are composed of surviving gas and dust protected by the shadows of these concentrations.

The Lives of the Stars_

In the early 1900s two astronomers – a Dane, Ejnar Hertzsprung, and an American, Henry Norris Russell – independently plotted stars on a chart that proved to be a road map to understanding stellar life cycles. They positioned the stars from top to bottom according to their true brightness (that is, after allowing for their differing distances from Earth), and from left to right according to their colour. Colour is an indication of temperature: cooler stars are more reddish; hotter ones, such as the Sun, are white; and the hottest ones are blue.

The stars fall into groups on the Hertzsprung-Russell (HR) diagram. Later theorists have been able to explain why stars appear at particular points in the HR diagram and how they move through it during the course of their lives.

⋯⋗ The diagonal band running across the diagram is called the main sequence. Stars spend most of their lives here, converting hydrogen into helium in nuclear reactions. Low-mass stars are cool and dim, at the bottom right of the main sequence. High-mass stars are hot and bright, at the top left.

⋯⋗ At the top right are huge, cool giants and supergiants.

⋯⋗ At the bottom left are faint but hot white dwarfs.

Something to Think About . . .

Each unit area of a hot star's surface is brighter than a unit area of a cooler star's surface. But a star's total brightness depends on its size and hence its total surface area. If two stars have the same temperature, the larger one will be brighter. If two stars are equally bright, the one that's cooler must be larger. These two facts mean that as you go up in the HR diagram (towards brighter stars) and to the right (towards cooler stars) you are moving towards larger stars.

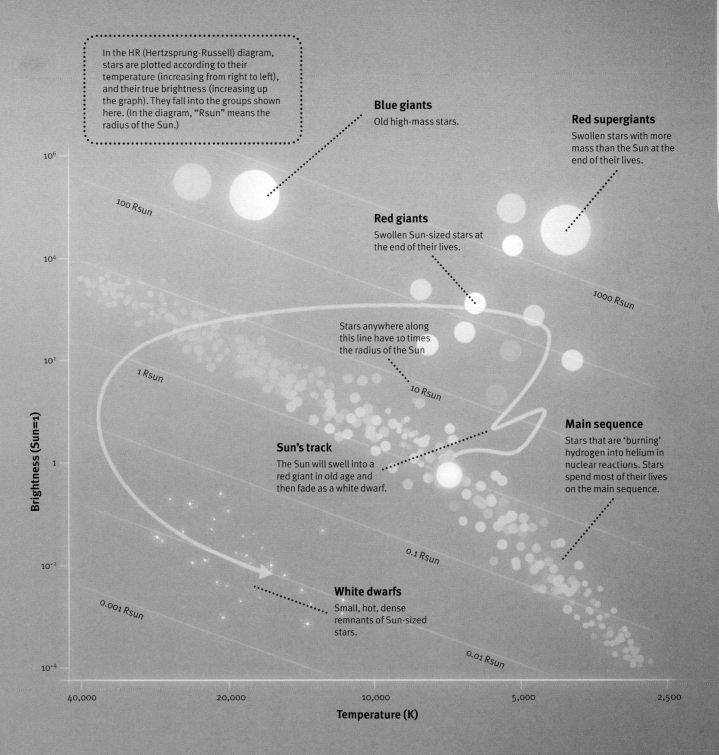

In the HR (Hertzsprung-Russell) diagram, stars are plotted according to their temperature (increasing from right to left), and their true brightness (increasing up the graph). They fall into the groups shown here. (In the diagram, "Rsun" means the radius of the Sun.)

Blue giants
Old high-mass stars.

Red supergiants
Swollen stars with more mass than the Sun at the end of their lives.

Red giants
Swollen Sun-sized stars at the end of their lives.

Stars anywhere along this line have 10 times the radius of the Sun

Main sequence
Stars that are 'burning' hydrogen into helium in nuclear reactions. Stars spend most of their lives on the main sequence.

Sun's track
The Sun will swell into a red giant in old age and then fade as a white dwarf.

White dwarfs
Small, hot, dense remnants of Sun-sized stars.

100 Rsun

1000 Rsun

1 Rsun

10 Rsun

0.1 Rsun

0.001 Rsun

0.01 Rsun

10^6

10^4

10^2

1

10^{-2}

10^{-4}

Brightness (Sun=1)

40,000   20,000   10,000   5,000   2,500

Temperature (K)

159

What Makes Stars Burn?_

The reactions in the heart of a star are nuclear – that is, they involve changes in the nuclei, or cores, of atoms. The ordinary burning of, say, petrol or coal, is a chemical reaction, involving the outer layers of atoms.

In the incredible heat and crushing pressure of the Sun's core, atoms are broken down, with positive nuclei and negative electrons flying around independently. Protons can collide to build up helium nuclei, releasing energy. Late in the Sun's life, when supplies of hydrogen run low, it will burn helium in its core to form carbon and oxygen. The Sun becomes unstable, pulsating in size and brightness.

Stars that are four or more times as massive as the Sun can burn carbon and oxygen to form heavier nuclei, and then burn these, all the way up to iron nuclei, which each have 26 protons and 28 neutrons. This marks a violent end for the massive star (see pages 162–163).

Atoms in most of the matter around us consist of a nucleus, which contains most of the mass of the atom and has a positive electric charge, and a surrounding cloud of electrons. The electrons have very little mass, and they have negative electrical charges that balance the positive charge of the nucleus. The nucleus is made of smaller positively charged particles called protons.

The simplest atom is that of hydrogen. Its nucleus is just a solitary proton and a single electron circles it.

The next-simplest atom is that of helium: there are two protons in the nucleus, circled by two electrons. The nucleus would be blown apart by the electrical repulsion between the two protons, except that there are also two other particles called neutrons. A neutron has almost exactly the same mass as a proton, but no electric charge.

Protons (hydrogen nuclei) collide in the heart of a star to build up heavier nuclei, which in turn collide to build still-heavier nuclei. There are many such processes; in the Sun, the main reaction is the one pictured here. Its net effect is that four protons are converted into a helium nucleus. Energy is produced in the form of gamma rays, which are short-wavelength radiation. Positrons (positively charged electrons) and neutrinos (ghost-like electrically neutral particles) are also emitted.

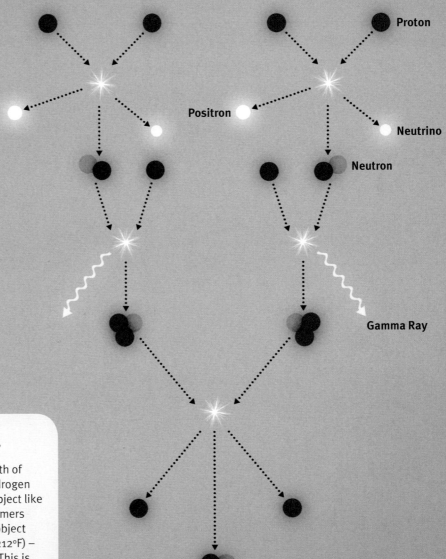

Proton

Positron

Neutrino

Neutron

Gamma Ray

Something to Think About...

The smallest possible star is one that has ¹/₁₄th of the Sun's mass. Within a smaller mass of hydrogen gas, nuclear reactions can never begin. An object like this is called a brown dwarf. In 2010, astronomers discovered the coolest brown dwarf yet: an object whose surface temperature is about 100°C (212°F) – about as hot as boiling water here on Earth. This is heat left over from the gravitational energy released when the brown dwarf formed.

Death of the Stars_

As long as the supply of hydrogen lasts, a star will move very little from its original position on the main sequence in the Hertzsprung-Russell (HR) diagram (see pages 158–159). The core will have become almost entirely helium, with some carbon and oxygen, and around it there will be just a thin shell in which hydrogen is continuing to be converted into helium. For the Sun this period will arrive five billion years from now and ten billion years after the Sun's birth.

The star swells, becoming a red giant and pulsating in and out. The outer layers of the star expand. Then some of the outer layers are ejected as part of an expanding shell. Astronomers call these globes of gas planetary nebulae because in a telescope they look like the discs of planets. At the centre of the nebula is the hot core of the star, containing most of its original mass compressed into a volume the size of the Earth. The temperature is around 10,000°C (18,000°F), making it white. This white dwarf shines only feebly because it does not generate energy. It will slowly cool over trillions of years.

This is what the end will be for stars like the Sun. There will be a very different end for a more massive star, which is hot and dense enough at its core to burn not only helium, but the heavier nuclei produced from the burning of helium: carbon, neon, oxygen, silicon and iron are produced in turn. A shell of each element is formed where the element is unburned, surrounding inner, hotter regions where burning continues.

Something to Think About . . .

When the Sun runs low on hydrogen fuel, our star will swell until it reaches to the present orbit of Mars. At the same time it will lose up to a third of its mass as it sheds its outer layers into space. It's therefore uncertain whether the Earth will spiral outwards or be swallowed up. But there is no doubt the Earth will burn up as the Sun grows hotter.

Key
- Hydrogen
- Helium
- Carbon
- Neon
- Oxygen
- Silicon
- Iron

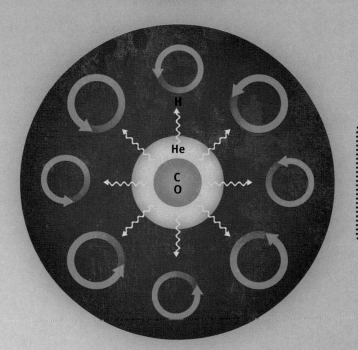

H
He
C
Ne
O
Si
Fe

**Massive star near
the end of its life**

The star has an 'onion' structure of
concentric shells. Each shell consists
of the products of a previous stage of
nuclear reactions. The final stage is the
formation of iron from silicon, which is
completed in minutes. Finally, there is
a colossal explosion that leaves
behind a super-dense remnant.

H
He
C O

**Solar-mass star near
the end of its life**

Hydrogen is nearly exhausted in the
hot, dense core and is being burned
only in a shell around it. The 'ash' of
past burning, consisting of helium,
carbon and oxygen, has accumulated
in the core.

Deadly Embrace_

Solitary stars end their lives either by shrinking quietly to dim white dwarfs or in violent supernova explosions. But many stars are binaries, travelling through space with partners, and their life story is more complicated.

Binary stars orbit each other happily for most of their lives while they are both on the main sequence (see pages 158–159), but one will be more massive, and will be the first to evolve towards the giant stage. As it swells, its grasp on its own outer layers becomes weaker than the gravitational pull of the partner star. Matter from the first star is drawn into a disc circling the second star, and then falls on to its surface.

The first star is reduced to its hot core (that is, to a white dwarf – see pages 154–155), while the second grows fat and enjoys a life extension. Eventually the second star ages and swells, and takes its turn to shed matter on to the first star. This dumping of fuel on to the white dwarf causes an enormous explosion, briefly billions of times brighter than the Sun, and the companion star may be flung out of the system. The remnant of the explosion, which may be a white dwarf or an even denser neutron star (see pages 168–169), may be visible as a faint point of light at the centre of a spectacular expanding gas cloud.

Something to Think About ...

The type of supernova resulting from stellar cannibalism is called a Type Ia supernova, and is even more violent than other types of supernova that mark the death of a solitary massive star. Their brightness is so great and so precise that they can be used as standards to measure the distance of remote galaxies when they are observed in them.

Stellar cannibalism

In a double star system, the more massive partner has evolved faster and has started to swell towards the red giant stage of its life. Its partner has started to swallow the larger star's outer layers. But the cannibal star itself will eventually grow old and start to swell, and then it will be the one having mass stolen by its companion star.

Seeds of New Planets_

The death of a massive star in a supernova explosion is just what the Universe needs to nourish a rising generation of new planetary systems. The star's innards, rich in heavy nuclei synthesized during its lifetime and others formed during the moments of the explosion, mingle with the interstellar gas and dust. These elements are mixed into the new stars and planets born from these clouds. In at least one location we know of, our own home, these ingredients have created life. It's not just a line from a song – we really are stardust.

As shock waves from the supernova travel through the interstellar gas clouds, they also lend a hand in initiating the process of star formation, by starting the process of collapse in regions where the gas is denser (see pages 100–101).

Something to Think About . . .

The great English cosmologist Fred Hoyle never believed that the Universe originated in a hot, dense state – in fact, he coined the derisive name 'Big Bang' for that theory. This meant he needed to find somewhere for the heavy nuclei to be 'cooked', and together with colleagues, he worked on the idea that heavy elements are built up in supernovae and scattered through space. At the same time, Big Bang theorists were struggling to cook the heavy elements in the Big Bang itself, but failed. The modern picture is that only helium (with two protons in its nucleus) and lithium (with three protons) were synthesized in the Big Bang: Hoyle and his colleagues were right about the rest.

The debris of a supernova explosion spreads across light years of space. The hydrogen within it glows pink. Mixed with the hydrogen are heavier nuclei synthesized in the core of the star before it exploded, and in the explosion itself. A fraction of this matter will go into the building of planetary systems.

06.9 Pulsars: Beacons in Space_

Stars can leave a variety of corpses behind when they die. We've seen that these can be white dwarfs, in which a mass about that of the Sun is crammed into a sphere the size of the Earth (see pages 162–163). But larger stars can collapse into something even more fantastically dense than a white dwarf: a neutron star. In a neutron-star nuclei and electrons have been crushed together to form a ball of neutrons as dense as an atomic nucleus. It packs a mass similar to that of the Sun into a sphere as wide as a city. If a cupful of a white dwarf were brought to Earth, it would weigh 300 tonnes. The same volume of matter from the centre of a neutron star would weigh 300 billion tonnes.

It was predicted in the 1930s that neutron stars should exist and would be formed in the collapse of supernovae. In the 1960s a few observations of suspected neutron stars were made. Then in 1967, a rapidly pulsating radio signal was picked up from the constellation Vulpecula, the Little Fox. It 'ticked' more regularly than the best atomic clocks. This pulsar was code-named LGM ('little green men') for a while, as its discoverers briefly toyed with the idea that it was a signal from an alien civilization. But there was no sign of a message and soon other pulsars were discovered in widely different parts of the sky.

A pulsar is actually a fast-spinning neutron star, with a hot spot on its surface that emits radiation strongly at radio wavelengths, and less strongly at visible and X-ray wavelengths. As it spins, the beam of radiation sweeps the sky and in the few cases where the beam hits the Earth, we see the signal. Thousands are now known.

Something to Think About . . .

The surface gravity of a neutron star is hundreds of billions of times the gravity we're used to here on Earth. Ordinary matter doesn't have the strength to resist that sort of force: if you fell on to the star you'd be squashed to the thickness of an atom.

Neutron star
A neutron star whirls rapidly on its axis. It has a mass similar to the Sun's, crushed into a sphere a few kilometres across. A disc of gas, the remnant of the parent star, girdles it. Jets of radiation and particles shoot out and are swept round the sky like lighthouse beams. If they sweep across the Earth, we see the neutron star as a pulsar.

06.10 Black Holes: a Star Vanishes_

Of all the final states a star can reach, the strangest is a black hole. Any star remnant that is greater than about three solar masses may end as a black hole. The black hole is formed when matter starts to collapse as nuclear fires run low and the process never stops. At least, according to all the physics we know, there is no end to its collapse. It dwindles to a point called a singularity. There are other ways in which black holes can form. (See pages 178, 190 and 192–193 for more on black holes.)

The fantastically strong gravity near the singularity distorts time and space. The singularity is surrounded by a small region whose surface is called the event horizon. Light or matter that passes through the event horizon can never escape, and as far as our knowledge of what's going on inside is concerned, the corpse of the star has vanished from the Universe.

But it continues to make itself felt – it still has its gravitational pull. The black hole may exist in a binary system with another star quite peacefully – until the ordinary star ages and swells, and matter may be pulled into its black hole companion, just as with other binary systems (see pages 164–165).

Something to Think About . . .

A black hole wandering alone through space, without a telltale companion or glowing disc of matter encircling it, can be detected, even though no matter or radiation is being emitted. The gravitational field of any object can affect light passing close by, rather like a lens. This lensing effect by a stellar-mass black hole can cause a background star apparently to brighten and shift temporarily. Telescopes making automated searches against the rich star fields of the central Milky Way have found thousands of such events.

A black hole bends the space and time around it, and it swallows light and matter. But on the very brink of disappearing, the in-falling matter is superheated, and intense jets of particles and radiation are ejected into space.

171

Exoplanets: New Worlds_

Among the hundreds of billions of stars in our Galaxy, how many have a family of planets? And how many of those planets harbour life? Every year we discover scattered examples of exoplanets (planets circling stars other than the Sun) and gain a better understanding of what the total number of planets might be.

A family of planets has been found circling a star called Kepler-11, named after the Kepler spacecraft, which discovered the system. The star is about as hot as the Sun, but five of the planets are closer to it than Mercury is to the Sun, and the outermost is only slighter further away than Mercury, so all the planets would be scorched. They are all larger than the Earth and are unlikely to support life.

The best candidates for planets capable of supporting life exist among those circling the red dwarf star Gliese 581. Again, all are several times the mass of the Earth. The larger a distant planet is, the easier it is to detect from the Earth. Exobiologists (scientists who search for life beyond the Earth) would like to find many planets of about Earth's mass circling in the habitable zones of their stars – that is, the distance ranges in which liquid water, essential to life as we know it, can exist.

Something to Think About . . .

The Kepler space telescope stares at one area of the sky, monitoring the brightness of over 100,000 stars. It can detect the dimming of a star equivalent to that caused by a fly crawling across a car headlight viewed from several kilometres away.

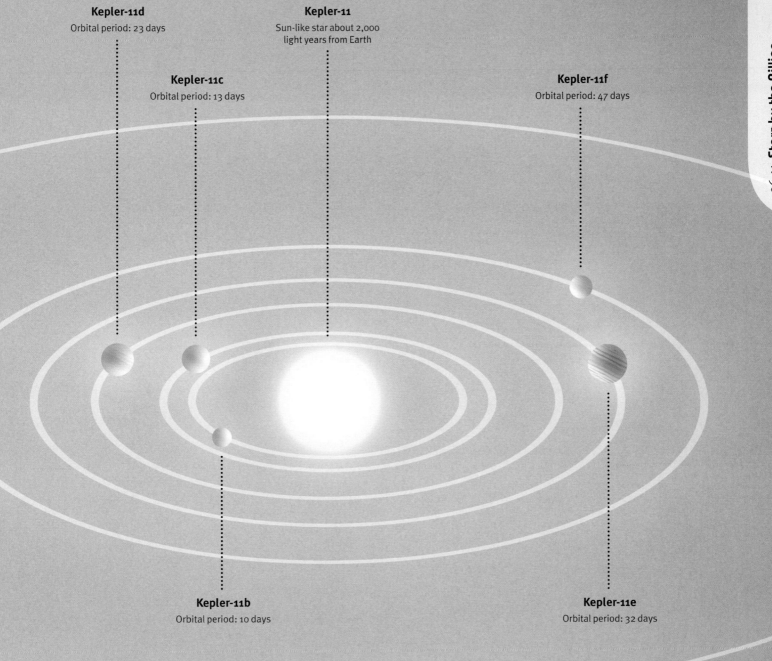

Kepler-11d
Orbital period: 23 days

Kepler-11
Sun-like star about 2,000
light years from Earth

Kepler-11c
Orbital period: 13 days

Kepler-11f
Orbital period: 47 days

Kepler-11b
Orbital period: 10 days

Kepler-11e
Orbital period: 32 days

06.12 Are We Alone?_

Discoveries of exoplanets come thick and fast: there could well prove to be billions in our Galaxy. Does that mean there are other intelligent races somewhere in our star system?

This depends on many factors, spelled out in the Drake equation (see right). Some factors are encouraging: the vast number of candidate planets, the speed with which life appeared on Earth (less than a billion years after the surface cooled), the fantastic extremes of environment in which life flourishes – from the cold of Antarctica to the scalding heat of submarine vents in the Pacific.

The Search for Extraterrestrial Intelligence (SETI) began in earnest in 1960 when a radio telescope scanned selected frequencies while pointing in turn at two nearby Sun-like stars, Tau Ceti and Epsilon Eridani. Today networks of radio telescopes scan the skies, and optical telescopes look for laser signals – so far without success.

Something to Think About . . .

The Italian-American physicist Enrico Fermi pointed out a problem with the idea that life is common in our Galaxy. If there are many planets containing life, with billions of years of development behind them, many should also have achieved interstellar flight. Where are they? There are no signs of alien visitors in the Solar System (UFO enthusiasts notwithstanding). Maybe there are insuperable barriers to interstellar travel; or life is so rare that we're the first to reach the level of interstellar communication; or they're hiding from us; or we are unique in the Galaxy …

Drake Equation

An equation used to estimate the number of civilizations in the Milky Way that we might be able to detect. The equation was devised in 1961 by Frank Drake at the University of California, Santa Cruz. All the terms in it have highly uncertain values.

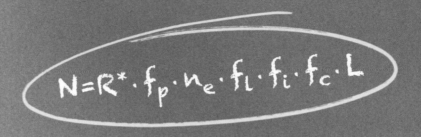

$$N = R^* \cdot f_p \cdot n_e \cdot f_l \cdot f_i \cdot f_c \cdot L$$

N = number of civilizations in our galaxy with which communication might be possible
R^* = number of stars formed per year in our galaxy
f_p = fraction of stars that have planets
n_e = average number of planets that could support life per star that has planets
f_l = fraction of potentially life-supporting planets that develop life

f_i = fraction of planets with life that develop intelligent life
f_c = fraction of civilizations that develop a technology detectable by us
L = length of time for which such civilizations are detectable

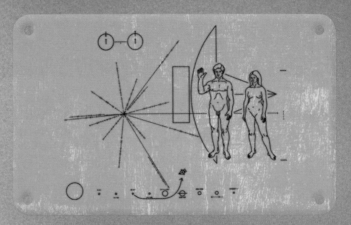

Pioneer Plaque

Gold-plated plaques on the Pioneer 10 and 11 spacecraft launched in the 1970s are addressed to any extraterrestrials that might intercept the craft. They show a man and woman and the location of the Earth in the Galaxy.

Arecibo message

A string of characters was transmitted into the void from the giant radio telescope at Arecibo, Puerto Rico, in 1974. Including information about the chemistry of life, the dimensions of the radio telescope, and the dimensions of human beings, the message was aimed at the globular star cluster M13, 25,000 light years away.

Chapter 07.0

Galaxies_

Anatomy of the Milky Way_

Our Solar System is located about halfway out from the centre of our home galaxy, the Milky Way. Almost everything that we can see in the sky belongs to this galaxy. The Milky Way – also called simply the Galaxy (with a capital 'G') – contains 200–400 billion stars. With the naked eye we can see only about 5,000 of these. You would have to travel thousands of light years to leave the Galaxy – hundreds of thousands of times further than any spacecraft has ever gone. This means we have no photographs of the Milky Way from the outside, but we can work out its shape from observations with telescopes, especially radio telescopes.

The Milky Way is just one galaxy in a group of galaxies called the Local Group. Another large spiral galaxy in the Local Group is the Andromeda Galaxy, which is very similar to the Milky Way, but larger and with several times as many stars. The Milky Way is moving through the Universe at about 600km (370 miles) per second.

Something to Think About . . .

There is a supermassive black hole (see pages 170–171) at the centre of the Milky Way. It apparently lies within a source of radio waves called Sagittarius A. The black hole has a mass equal to about four million Suns, packed into a volume not much larger than the Solar System.

Core of Milky Way Galaxy ·····················

Perseus arm ································

Centaurus arm ································

Sagittarius arm ································

Sun ································

Cygnus-Orion arm ································

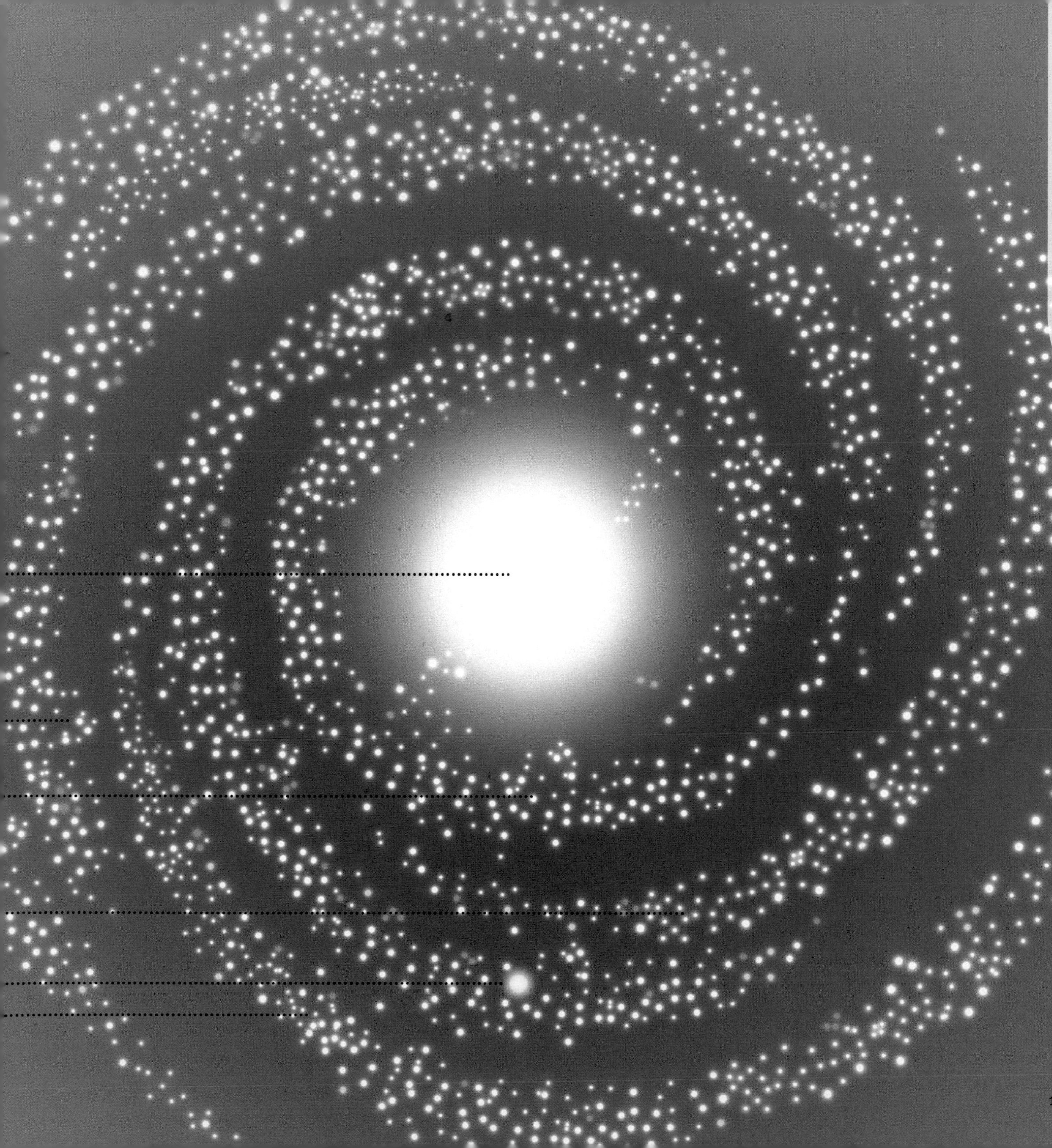

Cities of Stars_

Galaxies of all kinds are surrounded by swarms of tightly packed balls of stars, called *globular clusters*. These occupy a roughly spherical volume surrounding the galaxy called its *halo*. They were born when the galaxies were born and contain little gas and dust, so they now consist mostly of ageing, reddish stars. With perhaps several hundred thousand stars packed into a space only 100 light years across, the sky viewed from a planet near the heart of a globular cluster would be spectacular.

The clusters orbit the centre of their parent galaxy, plunging towards the centre and swinging outwards again. The cluster and the galaxy can be distorted by this encounter, although the individual stars within are too widely spaced to collide. It's possible that some of the most populous globular clusters in our neighbourhood, containing a million stars or more, are the cores of small galaxies that collided with the Milky Way and lost their outer regions to the larger galaxy.

Something to Think About . . .

Astronomers find globular clusters especially rewarding to study. Because all the stars are at practically the same distance from us, their apparent brightness is an accurate indication of their true brightness. And because they were nearly all formed at about the same time, the differences in their degree of development depend only on the differences in their initial masses.

The densely populated globular cluster M80 (meaning that it's number 80 on Charles Messier's 18th-century list of notable sky objects) lies about 30,000 light years from us. It was formed at the birth of the Milky Way and Sun-like stars within it have now evolved into red giants. There have been collisions (or perhaps just gentle mergers) between stars passing through its crowded central regions, rejuvenating those stars into apparently youthful *blue stragglers*.

Our Galactic Neighbours_

The Milky Way and its neighbour, the Andromeda Galaxy (see pages 62–63), are large spiral galaxies that are the two senior partners in a small cluster of galaxies, only about ten million light years across, called the Local Group. It contains a few dozen other small galaxies of all various types, which will be explained in more detail on the following pages.

The main components of the Local Group are:

The Milky Way Galaxy, consisting of up to 400 billion stars and with a mass about 700 million times that of the Sun. It has a bar-shaped nucleus.

The Andromeda Galaxy, consisting of perhaps a trillion stars, but probably having a mass about the same as the Milky Way. It too may have a bar-shaped nucleus.

M33, a spiral galaxy in the direction of the constellation Triangulum, the Triangle. M33 has a mass of about 50 billion Sun-masses and contains 40 billion stars.

The Larger Magellanic Cloud, an irregular galaxy, with a mass of about ten billion times that of the Sun.

The Smaller Magellanic Cloud, another irregular galaxy and a smaller twin to the Larger Magellanic Cloud, with a mass of about seven billion times that of our Sun.

The Canis Major Dwarf Galaxy, a tiny irregular galaxy that is the closest galaxy to the Earth, and is being torn apart by the Milky Way's gravity.

The Sagittarius Dwarf Elliptical Galaxy, orbiting at a distance of 50,000 light years from the core of the Milky Way.

Something to Think About . . .

On a large scale, all galaxies are rushing apart because of the expansion of the Universe. At short distances however, the speed of expansion is small compared with the random movements of galaxies within their own groups and clusters. In fact, the Andromeda Galaxy is approaching the Milky Way, and the two galaxies will collide in 4.5 billion years.

Andromeda Galaxy

Triangulum Galaxy

The Local Group

Our Milky Way galaxy belongs to a small cluster, the Local Group. The Andromeda Galaxy, 2.3 million light years away at the other side of the group, is virtually a twin spiral, and several dozen smaller galaxies of all types are fellow members.

Milky Way Galaxy

07.4 Spiral Galaxies_

Spiral galaxies show a beautiful structure that reflects different populations of stars.

··⇢ A spiral galaxy has a central bulge, which may be globular or bar-shaped. It consists of cooler, reddish, older stars.

··⇢ A relatively thin disc extends from the core of the galaxy, containing spiral arms. They consist of dark dust and gas. Embedded in these clouds are younger, bluer stars.

··⇢ A spherical halo of globular clusters (see pages 180–181) and individual stars envelops the disc of the galaxy. These are also reddish.

The origin of spiral structures in galaxies is not well understood. It seems to arise as a galaxy is built up by the merging of smaller ones. The mergers cause the galaxy to spin faster, and this causes the disc to form.

Something to Think About . . .

The arms of a spiral galaxy are like ripples – density waves passing through the gas and dust of the galaxy. Where the gas and dust are concentrated by the wave, there is a starburst of stellar formation, and young blue stars stud the dark clouds. The density waves move on and leave stars behind, to be overtaken by the next arm.

Some spiral galaxies are loosely coiled and have only two arms. Others are more tightly coiled and can have more arms that create a spectacular pinwheel appearance.

Young blue stars

Bulge

Old red stars

Dust and gas clouds

Galactic Footballs_

Elliptical galaxies are generally shaped like rugby balls or American footballs, although they can have a range of shapes, varying from very elongated to spherical. They are rather like the central bulges of spiral galaxies: they glow reddish because their stars, which are generally of low mass, were born in the early days of the Universe and have evolved to the red-giant stage. There is little interstellar gas and dust, so young stars are rare. Elliptical galaxies don't rotate – the stars orbit the centre individually – and they can be huge: the largest have over a trillion stars.

Elliptical galaxies seem to be the product of galactic greed, having grown by the mergers of smaller galaxies. The result of such a collision will often be an elliptical galaxy. The bulky products of collisions continue to swallow smaller galaxies. (See pages 190–191.)

Something to Think About . . .

Elliptical galaxies seem to have been the first to form – they are more common among very distant galaxies, which we see as they were in the first few billion years of the Universe. Some galaxies that were originally elliptical developed into spirals, which are more common in the nearby Universe.

An elliptical galaxy glows reddish because it consists mostly of aged, cool, giant stars. There is little gas and dust from which new stars, which would be hotter and bluer, could form.

187

Shapeless Galaxies_

Many galaxies can't be classed as spiral or elliptical, but instead are irregular in shape. These are small, dusty, bluish collections of stars. Sometimes they are distorted by some extreme activity occurring in the core (see pages 192–193). Or they may have lost their shape through collisions or close interactions with other galaxies (see pages 190–191).

As astronomical techniques have improved, it has become possible to detect subtle traces of the past structures of irregular galaxies, and they are then reclassified. One of the most famous galaxies formerly regarded as irregular is the Cigar Nebula, or M82, in the constellation of Ursa Major, the Great Bear. It is a cigar-shaped mass of stars and dust. At its core, new stars are being formed at ten times the rate they are being formed in the whole of the Milky Way. We now know that two spiral galaxies are colliding here but the spiral arms are hard to see.

Something to Think About . . .

Two of the Milky Way's closest neighbours are the Large and Small Magellanic Clouds, in the southern skies. They show traces of barred-spiral structure (see pages 184–185). The galaxies have apparently been disrupted by interactions with each other and with the Milky Way. A bridge of gas and stars links the two smaller galaxies, and a stream of matter curves around the outside of the Milky Way, while the disc of the Milky Way is warped by the Clouds.

Irregular galaxies have no definite shape, contain a large amount of gas and dust, and many young, bright blue stars.

Galaxies in Collision_

Vast though the spaces between galaxies are, collisions between galaxies are surprisingly common, and they were much more common when the Universe was younger and more crowded. Many of these celestial train wrecks can be observed, and many more galaxies show traces of past close encounters.

When galaxies collide, the stars don't hit one another – proportionally, stars are much further apart in relation to their size than galaxies are. But the gas clouds in the galaxies do collide and settle at the centre of the merged galaxy. There is a burst of star formation here, and a supermassive black hole may form.

More commonly the two galaxies will narrowly avoid a collision, and instead they will pull stars and other matter out of one another in long streams.

Something to Think About . . .

The Cartwheel Galaxy exhibits the results of a galactic collision. Half as broad again as the Milky Way, it takes the form of a rim of bright young stars and powerful X-ray sources, around the edge of a disc-shaped galaxy with weak 'spokes'. We're apparently seeing it 200 million years after a smaller galaxy dived straight through the disc of what was then an ordinary spiral galaxy. A shock wave of star formation rippled outwards, forming the bright ring. The spokes are actually spiral arms re-establishing themselves.

Galaxies heading for a smash: the
Mice Galaxies, 290 million light years
away from us in the constellation
Coma Berenices, are approaching a
collision. The long 'tail' stretching
from one is a stream of stars ejected
by the gravitational field of the other.
The two galaxies have probably
collided several times in the past.

07.8 Violent Galaxies_

In 1963, a faint, star-like object was found to be at the exact location of a radio source called 3C 273. The spectrum of the 'star' was mysterious until it was realized that the lines in the spectrum had been massively shifted toward the red (see pages 64–65). This meant that it was at cosmic distances – several billion light years away – and was actually incredibly 'bright' at radio wavelengths. The tiny object was apparently pouring out the energy of 100 galaxies like the Milky Way. It was christened a quasar, for 'quasi-stellar radio source'.

More and more quasars were discovered, as well as equivalent objects that poured out unbelievable amounts of energy in the optical wavelengths. These distant, intensely bright objects are now generally called active galaxies.

These fantastic power levels were difficult for the scientific world to accept, until the idea of matter falling into black holes and vanishing from the Universe forever was developed. (See pages 170–171.) In the heart of an active galaxy, stars, dust and gas circle the black hole just before being swallowed. When the matter at last falls in, a full ten per cent of its mass can be converted into pure energy – a far more efficient power-generation process than any other known.

Something to Think About . . .

Active galaxies are far more frequent at great distances – that is, in the first few billion years of the Universe. This suggests that all galaxies have a violent youth, which dies down as the black holes in their centres run out of matter to swallow.

Jets of high-energy particles blast from the heart of an active galaxy, pouring out radio waves, X-rays and visible light that can be detected far across the Universe. A black hole at the centre of the galaxy is responsible.

07.9 Clusters of Galaxies_

Galaxies are generally sociable – they form clusters. The clusters, which may include smaller groupings such as our Local Group, measure up to about 30 million light years across, and may contain thousands of galaxies. The clusters themselves form larger groupings – superclusters – up to perhaps 300 million light years across. Our Local Group is not part of a cluster, but along with hundreds of other groups and clusters, forms part of the Virgo supercluster measuring about 110 million light years across.

Clusters of galaxies were once called 'metagalaxies', and are rather like scaled-up galaxies:

⇢ Some are regular, or spherical in shape, with numbers of galaxies falling off smoothly from a concentrated centre.

⇢ Some are irregular.

⇢ Sometimes they collide with one another, just as some galaxies do.

All show, by their gravitational effects, that they have a great deal more mass than is accounted for by the visible and non-visible radiation we can observe (see pages 206–207).

Something to Think About . . .

The poet William Blake saw 'a world in a grain of sand'. Astronomers can do almost as well: if you hold a grain of sand up to the sky at arm's length, the area of sky it covers contains 10,000 galaxies.

In a cluster of galaxies, there may be thousands of galaxies, of all types and each containing tens or hundreds of billions of stars.

There is invisible gas between the galaxies in a cluster. It gives out X-rays that show the gas is at tens or hundreds of millions of degrees. The mass of the gas in a cluster is typically twice as great as the mass of the visible galaxies.

Walls & Voids_

As we look out across the Universe at its very largest scale – billions of light years – and therefore look backwards in time, its structure shows clearer and clearer traces of the processes that took place in its youth.

The grouping of matter doesn't end with superclusters of galaxies. These in turn group into walls and filaments on a scale of hundreds of millions of light years. These surround great spaces, almost empty of galaxies, called voids, typically about 75 million light years across. The Universe at this scale is 'foamy', with the voids dominating and the galaxy clusters forming the voids' walls: it is more like a mass of soap bubbles than a Swiss cheese.

This structure probably originated when the Universe was about 300,000 years old, and cooled below a temperature of a few thousand degrees. Electrons and nuclei of hydrogen and helium that had emerged from the Big Bang combined to form atoms. Fluctuations in the density of the hot matter that had developed before that moment then became 'frozen in'. Regions of higher density attracted more matter, which accumulated into galaxies. The Universe has expanded a thousandfold since then, and these early variations have expanded into the structures we see today.

Something to Think About . . .

The 'clumping' of matter comes to an end eventually. On a scale greater than about 300 million light years, the groupings of galaxies are scattered evenly through space – there are no 'super-superclusters'. Cosmologists call this balancing out of irregularities the 'End of Greatness'.

The large-scale structure of the Universe. At this scale, spanning a billion light years, individual galaxies and even clusters of galaxies can't be seen: they merge into luminous tendrils and sheets enfolding huge dark voids, within which there are very few galaxies.

Chapter 08.0

The Cosmos_

Stretching Space_

If the Universe expanded from a huge explosion 13.5 billion years ago, where did the explosion happen? And what was going on outside that first pinpoint of matter as it exploded?

The answer to the first question is: it happened nowhere – or everywhere. There is no single place in the Universe we can point to today and say that the explosion happened there.

And the answer to the second question is: there was no 'outside'. In the first moments when the Universe was the size of a football it filled all the space that existed. And when after a year, it had grown to the size of a present-day galaxy, it still filled all the space there was.

Think of the growing Universe as an expanding balloon – but space is not the interior of the balloon, it is the elastic skin of the balloon. The skin provides a two-dimensional analogy to our ordinary three-dimensional space. If you were a two-dimensional ant inhabiting the surface of the balloon, for you there would be nothing outside it – but there would be a definite size to your universe, a size that increased as the universe expanded.

Something to Think About . . .

Galaxies and galaxy clusters don't get bigger with the general expansion of the Universe – gravity is strong enough to hold them together. Bigger structures – superclusters and the walls and filaments they are grouped into (see pages 196–197) – do expand at the same time as they move further apart.

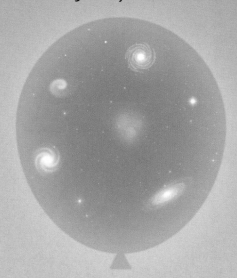

3 billion years old

Today

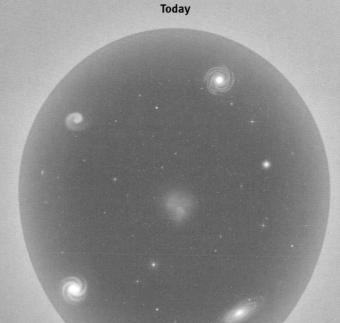

7 billion years old

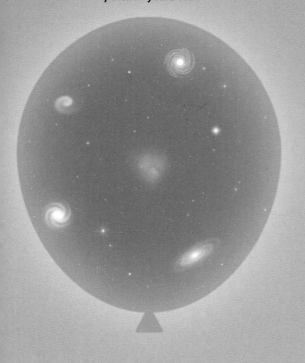

A toy balloon models the expanding
Universe. The model galaxies attached to
the surface move further apart. Each one
'sees' all the others as moving away from it.
From the galaxies' point of view, no galaxy
is special – no point in the surface of the
balloon is the centre of the expansion.

A Brief History of the Big Bang_

Cosmologists are sure they know the story of the development of the Universe from the Big Bang to the present. Their confidence is largely based on the fact that the early Universe was very hot, and the behaviour of matter is simpler in some ways at high temperatures. But in the earliest fractions of a second, it was too darned hot. The temperature of the Universe was so unimaginably high that our theories of physics break down completely and we can say nothing with certainty about what was going on.

Timeline of the Universe

Here are some milestones on the Universe's path to the present, expanding and cooling all the way.

Note: here compact scientific notation is used for very large and very small numbers. For example:

··→ 10^6 means six tens multiplied together, or 1,000,000. So 10^{35} means the number 1 followed by 35 zeros.

··→ Also, 10^{-6} means 1 divided by 10^6 – a millionth – and 10^{-35} means 1 divided by 10^{35}.

The beginning
Nothing known – present physics not valid.

10^{-43} seconds
There are two fundamental forces: gravitation and a grand unified force.

10^{-35} seconds
The Universe inflates – blowing up from a size less than that of a subatomic particle to the size of a house in an instant.

10^{-32} seconds
The Universe expands at a more gentle pace, and the rate of expansion slowly decreases. The universe is a soup of particles and radiation. The fundamental forces have split into the four we know today.

10^{-6} seconds
Protons and neutrons are formed as more fundamental particles combine.

3 minutes
Some protons combine with neutrons to form helium nuclei and some heavier elements.

370,000 years
Electrons and nuclei combine to form atoms. Radiation travels freely and today forms the cosmic microwave background.

200 million years
First stars begin to shine.

1 billion years
First galaxies form.

9 billion years
The expansion of the Universe begins to accelerate.

13.7 billion years
The Universe today.

Ages of the Universe:
from the Big Bang
to today

13.7 billion years:
The Universe today

1 billion years:
First galaxies

370,000 years:
Atoms form

10^{-35} seconds:
Inflation

Big Bang

Something to Think About . . .

There are four fundamental forces in today's Universe:
gravitation, electromagnetism and two types of nuclear
force. These are very different in strength. But at the
beginning of the Universe, there was one super-force that
affected all particles equally, and a fraction of a second
after the Big Bang it split into the forces we know today.

The Fog at the End of the Universe_

All around the sky, as a backdrop to the remotest galaxies, is a cosmic fog, a steady hiss of radiation. It consists of a mix of wavelengths peaking at around 2mm ($\frac{1}{13}$in), about one-sixtieth of the wavelength used in a microwave oven.

All objects emit heat radiation corresponding to their temperature. This cosmic microwave background (CMB) radiation has the characteristic mix of wavelengths of an object at a temperature of just 2.7°C (5°F) above absolute zero. (Absolute zero is the lowest possible temperature, at which all the atoms in a body come to rest.) The temperature of the CMB is the average temperature of the Universe.

The CMB is radiation generated in the Universe about 370,000 years after its birth – mere moments on the cosmic timescale. Up to this time the Universe consisted of electrons, atomic nuclei and radiation all jostling one another. As this mixture cooled to about 3,000°C (about 5,400°F) electrons and nuclei combined to form neutral (electrically uncharged) atoms. These hardly interact with electromagnetic radiation, meaning that suddenly the Universe became transparent and radiation could travel almost unhindered across it. It's still travelling – and it's what we see as the CMB. Just as the Universe has been expanding and cooling all that time, the background radiation has stretched and cooled to its present low temperature.

Something to Think About . . .

You don't need a radio telescope to detect the CMB. It contributes a few per cent of the 'snow' that filled the screens of old-fashioned television sets when they were untuned.

Cosmic microwave radiation (CMB) forms a backdrop that lies beyond the furthest galaxies. It has travelled to us from when the Universe was just 300,000 years old. Here it has been colour-coded: red is hottest, dark blue is coolest – but the variations are only 1/30,000th of a degree at most. These ripples in the CMB represent ripples in the Universe when it began – ripples that were the seeds from which galaxies grew.

08.4 The Dark Universe_

Astronomers have recently come to suspect that everything that we can observe in the Universe – whether via light, radio, X-rays, neutrinos or whatever – may be only the tip of the cosmic iceberg. There are clear signs that there is far more matter in the Universe than meets the eye.

Galaxies rotate faster than the amount of matter in them would suggest. They also orbit in their clusters faster than the numbers of other galaxies suggest they should. The clusters would have broken up long ago unless their members are being influenced by something unseen – some 'dark matter'. (Not to be confused with 'dark energy' – see pages 208–209.)

There have been many candidates for what dark matter could be, including:

⋯⋗ MACHOs – Massive Compact Halo Objects, i.e. stars and very dense objects too faint to be seen, such as dwarf stars, black holes and neutron stars. However, these seem to account for only a small part of dark matter.

⋯⋗ WIMPs – Weakly Interacting Massive Particles, i.e. some undiscovered type of fundamental particles that interact through gravitation and the weak nuclear force (which is involved in radioactivity), but not through electromagnetism or the strong nuclear force (which holds nuclei together). There are experiments under way at the Large Hadron Collider near Geneva, Switzerland, as well as at other laboratories, to search for WIMPs.

Something to Think About . . .

An Italian research team called DAMA (for 'DArk MAtter') claims to have detected a WIMP wind blowing past the Earth. Their underground detector records hits from unidentified particles at a rate that varies through the year, peaking at the beginning of June. They claim this is due to the Earth travelling at different speeds through the Galaxy as it circles the Sun. But they're a lone voice – no other laboratory has yet detected such particles.

The abyss beneath

We observe the 'normal' matter and energy of the Universe with a wide range of radiation. But there's more than five times as much 'dark' matter that shows itself only by its gravitational pull. And beyond both of these there is three times as much of the utterly mysterious 'dark energy' that is driving the expansion of the Universe ever faster.

Dark energy
74 per cent

Dark matter
22 per cent

Normal matter and radiation
4 per cent

08.5 The Accelerating Universe_

An astonishing discovery was made in 1998. A survey of Type Ia supernovae (the sort that explodes after 'gobbling' the outer layers of a companion star) had enabled the distances of galaxies to be calculated to further out than ever before. When compared to their red-shifts, however, it appeared that the expansion of the Universe was not proceeding steadily. It wouldn't have been surprising if it were slowing, owing to the mutual gravitational attraction of all the matter in the Universe, but in fact it is speeding up.

Cosmologists are baffled by this apparent acceleration: they describe it as being due to dark energy appearing throughout space as the Universe expands. There are some tentative theories as to what dark energy might be, but none is well established.

When the Universe was smaller, there was less dark energy and less repulsion. Because of the gravitational attraction between objects, expansion actually began to slow down. But about five billion years ago the Universe had grown large enough for dark energy to overcome gravity. If dark energy continues to act in this way, the Universe will never stop expanding (see pages 216–217).

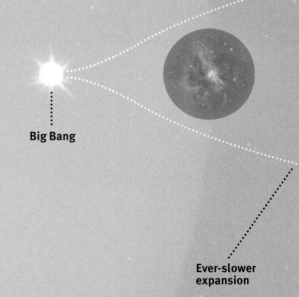

Big Bang

Ever-slower expansion

Something to Think About . . .

When Albert Einstein tried to model the Universe using his General Theory of Relativity, he assumed the Universe was neither collapsing nor expanding. To keep it in balance he invented a cosmological constant representing a force of repulsion between all particles of matter. When Hubble discovered the galaxies were rushing apart, Einstein described the cosmological constant as the greatest mistake of his life. But now it seems it was no mistake at all – the cosmological constant would have much the same effect that dark energy seems to have.

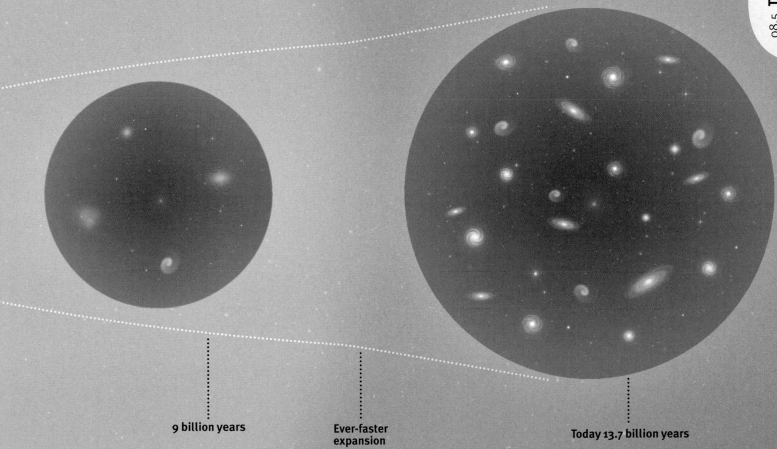

After the incredibly rapid expansion of the inflation era, the Universe expanded at a rate that was more sedate. However, after about nine billion years expansion began to accelerate once again.

9 billion years

Ever-faster expansion

Today 13.7 billion years

08.6 Why is the Sky Dark?_

In 1823 a German astronomer, Heinrich Olbers, discussed a problem that had intrigued others before him: why is the sky dark at night? If the Universe was infinite, eternal and unchanging, in whichever direction we looked, we should see a star: the whole sky should be bright.

Some astronomers believed the Universe was infinite – it was just that distant stars were so faint that they didn't contribute much to the brightness of the sky. However, their logic was flawed (see caption, opposite).

Olbers thought a very slight haziness in space, due to interstellar matter, would be sufficient to explain why the sky was dark. But later scientists realized this couldn't be right: any medium that absorbed starlight would heat up and begin to shine – resulting in a bright sky once again.

Then the stars – now known to be grouped into galaxies – were discovered to be moving away from us, and astronomers realized that this would stretch and weaken the light waves reaching Earth. But this isn't the whole answer: it has been calculated that if all the galaxies came to an abrupt halt, the amount of light would merely double – still leaving the sky dark.

The real solution to Olbers' problem is that we can see only as far as a cosmic 'horizon'. At a distance of 14 billion light years, all galaxies recede at the speed of light, and are invisible – as are those beyond that. And the number of stars inside that horizon is far short of what's needed to make the sky bright.

Something to Think About . . .

If we lived in an eternal, infinite and static universe, the distance to which stars would have to be visible to wallpaper over the sky would be staggeringly huge: using modern data for the spacing and brightness of stars, we'd have to look 10^{23} (100 billion trillion) light years into space. Compare that with the less than 14 billion light years we can actually see.

Further stars

Nearer stars

Earth

Distant stars are faint – but that's not the reason they don't make the sky bright. Distant stars should contribute just as much brightness to the night sky as nearby ones. The outer 'shell' of stars here is twice as far away from us as the inner shell, so each star looks only a quarter as bright from Earth. But the outer shell has four times as much surface as the inner shell – so the total light from the outer shell is the same as the total from the inner one. Taking more and more shells, the amount of starlight falling on the Earth builds up without limit.

08.7 From Universe to Multiverse_

In a Universe that has existed for less than 14 billion years, we cannot see further than 14 billion light years into space (see pages 210–211): that is as far as a light ray could have travelled in that time, and nothing is faster than a light-ray. Nothing beyond that distance can communicate with us or affect us. So it might seem that what lies within the cosmic horizon (the horizon lies just beyond the CMB – see pages 204–205) is effectively our whole Universe.

But many models suggest that the Universe we know grew from one bubble in a foam of bubbles at the first moment of the Big Bang. Neighbouring bubbles grew to universes beyond our cosmic horizon. They could live out their own life histories unknown to us.

The laws of physics could be different in different bubbles within this multiverse. If gravity were a little weaker, or the rate of inflation were a little faster, in one of these universes, the universe would expand too fast and the gas filling space would never collect into stars and planets. That universe would remain a sea of cold gas for ever.

If the balance of forces were the other way, the whole universe might collapse a moment after the Big Bang. Or the universe might be long-lived, but nuclear reactions might run faster, so that stars would form and burn only for millions rather than billions of years, and life would have no chance to develop.

Something to Think About . . .

Some serious cosmologists have been impressed by the way in which various features of our Universe are finely balanced – the strengths of gravitation and nuclear forces, for example – making it possible for stars to ignite and live for long periods and provide potential homes for life. But if there are an unimaginably enormous number of universes, with physical details varying from one to the next, then one of them would have to have just the right combination of properties to be a home for life. The existence of multiple universes could provide an explanation for 'fine-tuning' without the need for a supreme tuner.

The Big Bang might have created universes alongside our own – some rich in stars and life, some empty, some short-lived. The search for signs of them, in the form of hot and cold spots in the CMB, has already begun.

Before the Beginning_

Big questions present themselves where the Big Bang is concerned: what caused it, and what came before? (Of course, whatever cause we come up with, we'll still be able to ask the same question again about that. But explaining things is an addiction for scientists, and they won't give up the habit just because there's no end in sight.)

One answer is that both questions are meaningless, because time itself began with the universal expansion – there is no 'before'.

But according to other theories, what came before the beginning is that a universe like ours collapsed, formed a pinpoint-sized fireball and then exploded; and our present Universe will repeat the cycle, for ever (see pages 216–217).

Yet more theories suggest that beyond our familiar three dimensions of space and one of time, there are further dimensions that we can't perceive – though high-energy experiments with machines like the Large Hadron Collider might be able to explore them. Many three-dimensional embryo universes float through these higher dimensions. When two of these collided, our Universe began its history in the Big Bang. And when our Universe has faded, it could be re-ignited again and again in future collisions.

Previous universe

Something to Think About . . .

In 2010 cosmologists claimed that circular patterns they'd found in the CMB (see pages 204–205) were echoes of collisions between supermassive black holes in a universe that preceded ours. The idea was quickly rubbished by their colleagues – but it shows how in principle we could probe the era before the Big Bang.

Perhaps our Universe was born from the death of a previous one. The collapse of that older universe might look like the Big Bang, run in reverse. Galaxies would rush together, be consumed in a fireball, and be reborn in the Big Bang – this time, running forwards.

Our Universe

Collapse of previous universe

Expansion of our Universe

10^{-35} seconds:
Inflation

370,000 years:
Atoms form

1 billion years:
First galaxies

13.7 billion years:
Today

Things to Come_

A century ago it seemed clear how the Universe must end: with the lights going out as the stars' fires ran out of fuel, and the Universe subsiding into a darkness populated by the stars' cold corpses. But modern physics and cosmology offer a range of possible endings.

---> In the Big Crunch theory, the expansion of the Universe will one day slow and then reverse. In as little as 20 billion years, all the matter in the Universe could be falling inwards. As matter and radiation pack together more densely, the temperature will rise. The Universe will once again become a sea of charged particles and radiation. Gravity will become more and more intense and finally suck in the whole Universe.

---> A Big Crunch might lead to a Big Bounce as a new Big Bang flings matter out again to repeat the cosmic cycle. (But perhaps not – perhaps the Universe only gets one go at the sequence from Big Bang to Big Crunch.)

---> The Big Rip. It has been suggested that the expansion we see today will accelerate to a point where galaxies, then planetary systems, and then stars, planets and finally atoms would be torn apart.

---> The Big Freeze. The expansion may simply continue for ever, as its component stars flicker out.

Something to Think About . . .

If the Universe lasts for ever, it will pass a series of milestones – or gravestones – on the way:

1 trillion (10^{12}) years: Last stars formed.

2 trillion years: Galaxies beyond the Virgo Supercluster (to which our Local Group belongs) pass out of view.

10 trillion years: The long-lived red dwarfs cease to shine.

10^{34} years: Matter is locked up in white dwarfs, neutron stars and black holes; protons decay into lighter particles.

10^{43} years: All particles have been swallowed by black holes, now the sole denizens of the Universe.

10^{100} years: Black holes evaporate; the Universe is a sea of electrons, neutrinos and very low-energy photons.

Tomorrow and tomorrow and tomorrow ... according to one group of theories, the expansion of the Universe will one day stop and reverse, and the Universe will eventually be swallowed up in a Big Crunch. That may – or may not – mark the beginning of yet another journey from Big Bang to Big Crunch.

The Universe expands for ever

EXPANSION

The Universe starts to contract

Big Bang

Big Crunch

TIME

217

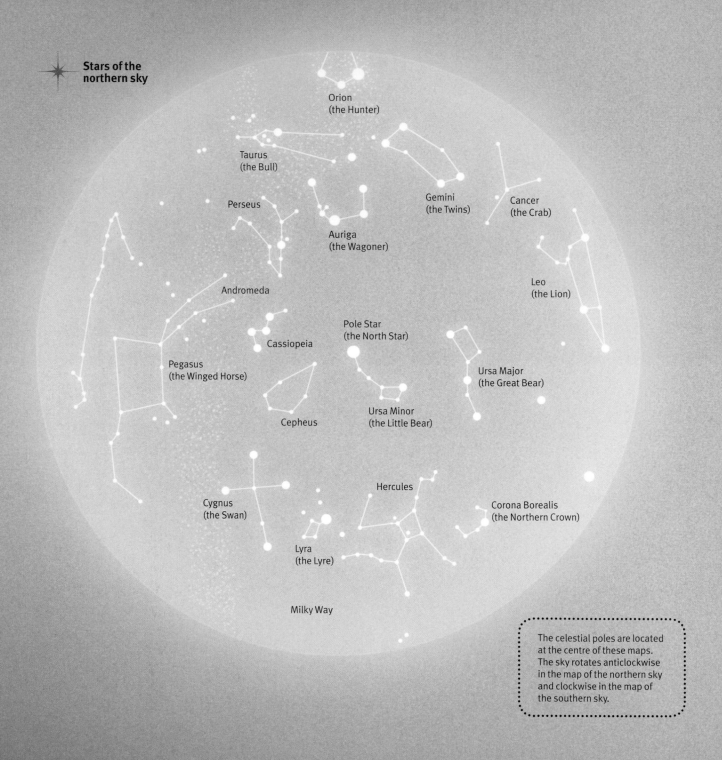

Stars of the
northern sky

Orion
(the Hunter)

Taurus
(the Bull)

Gemini
(the Twins)

Cancer
(the Crab)

Perseus

Auriga
(the Wagoner)

Leo
(the Lion)

Andromeda

Pole Star
(the North Star)

Cassiopeia

Pegasus
(the Winged Horse)

Ursa Major
(the Great Bear)

Cepheus

Ursa Minor
(the Little Bear)

Hercules

Cygnus
(the Swan)

Corona Borealis
(the Northern Crown)

Lyra
(the Lyre)

Milky Way

The celestial poles are located
at the centre of these maps.
The sky rotates anticlockwise
in the map of the northern sky
and clockwise in the map of
the southern sky.

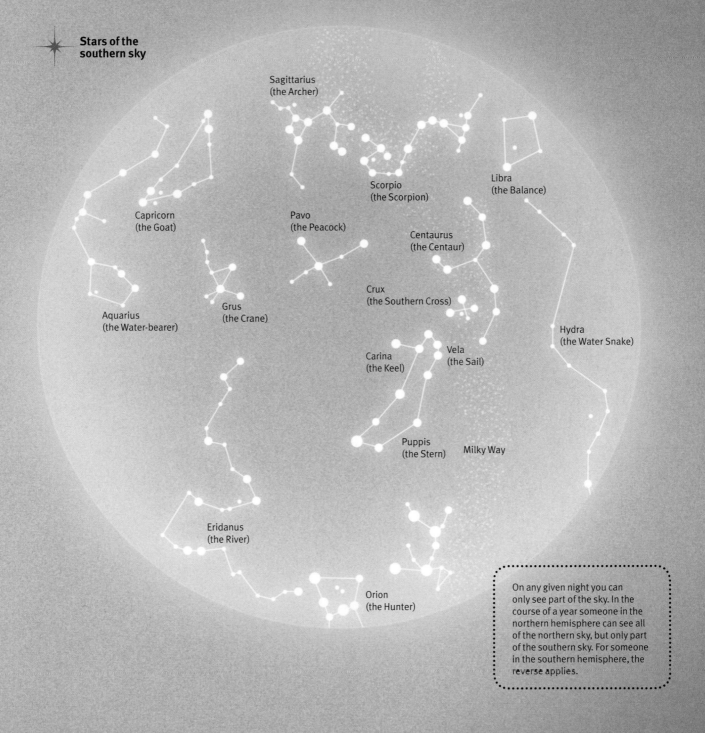

Stars of the southern sky

Sagittarius (the Archer)

Scorpio (the Scorpion)

Libra (the Balance)

Capricorn (the Goat)

Pavo (the Peacock)

Centaurus (the Centaur)

Aquarius (the Water-bearer)

Grus (the Crane)

Crux (the Southern Cross)

Hydra (the Water Snake)

Carina (the Keel)

Vela (the Sail)

Puppis (the Stern)

Milky Way

Eridanus (the River)

Orion (the Hunter)

On any given night you can only see part of the sky. In the course of a year someone in the northern hemisphere can see all of the northern sky, but only part of the southern sky. For someone in the southern hemisphere, the reverse applies.

Index_

Index_ *cont.*

Index_ *cont.*

About the author

Chris Cooper is a science writer and editor. His previous books include *The Solar System, How Everyday Things Work* and *Matter*. He has also edited and contributed to many books and encyclopedias, including *The Practical Astronomer, The Search for Infinity: from Quarks to the Cosmos, The Natural History of the Universe* and *Secrets of the Universe*. He lives with his family in the English country town of Bedford.